Lakes : Water
and Sediment Geochemistry

Lakes : Water and Sediment Geochemistry

BRIJRAJ KRISHNA DAS
Centre of Advanced Study in Geology
Panjab University, Chandigarh (India)

2008

SSPH

SATISH SERIAL PUBLISHING HOUSE
403, Express Tower, Commercial Complex, Azadpur, Delhi-110033 (India)
Phone : 011-27672469, 27672852, Fax : 91-11-27672046
E-mail :info@satishserial.com, hkjain1975@yahoo.com
Website : www.satishserial.com

Published by :

SATISH SERIAL PUBLISHING HOUSE
403, Express Tower, Commercial Complex, Azadpur, Delhi - 110033 (INDIA)
Phone : 011-27672852 Fax : 91-11-27672046
E-mail : info@satishserial.com, hkjain1975@yahoo.com

ISBN 81-89304-54-2

Printed at
SALASAR IMAGING SYSTEMS, DELHI-35

PREFACE

Water is essential for the life. Various uses of water such as drinking, irrigation, fisheries, industries and transport are made by humans, and humans are source of polluting water through sewage discharge, untreated effluent discharge, agriculture, deforestation etc. which also affect aquatic life to a large extent.

Natural process of weathering i.e breaking of rocks, their disintegration and transport, releases elemental ions to the water, rivers, lakes, oceans and sub-surface water. Rain water in this respect, plays a vital role as it interacts with the atmospheric CO_2 dissolving it (H_2CO_3) making water more reactive which attacks rocks exposed since millions of years to the atmosphere. This is the natural source of ionic concentration in waters. The geochemistry of natural waters has been discussed in great details by Drever, Berner and Berner, Stumm and Morgan and others.

The present book aims in discussing water and sediment geochemistry of lakes in the background of natural processes in respect of Indian lakes, the information and data in respect of these are lacking which may be of interest to limnologists, geochemists etc. Moreover, sediment geochemistry of lakes has perhaps not received much attention of the scientists and hence no recent published account is available. In view of this, a chapter on sediment chemistry with examples of investigated areas and data are included in this book. All examples discussed in this book are based on original data of the author who has been working since more than a decade and a half in this field. Most of the lakes investigated by him are from the Himalaya and some are from Rajasthan, Udaipur, and these are fresh water lakes. Data on coastal lakes could not be obtained hence not incorporated in this book. The Himalayan lakes are important because they are located at different altitudes which involves variation in climate and population. Some of these lakes, therefore, have very less anthropogenic interference and maintain their pristine character. The chemistry of such lakes, therefore, presents a real picture on natural processes that have occurred during the life of the lake system.

In order to make water and sediment geochemistry of lakes understandable, first few chapters of this book are devoted to general principles and chemical reactions etc. which are related to lake dynamics described by other authors and taken from them. The later chapters deal with examples of specific lakes encompassing original data on water and sediment geochemistry of lakes and their possible source. The last chapter introduces bio-geochemistry as an important tool for reconstruction of paleoclimate on the basis of lake sediment study and few examples of Himalayan lakes are cited. The book, it is hoped, will be useful to the students who have elementary knowledge of science as well as to limnologists, sedimentary geochemists, environmental geochemists, engineers, researchers and government organizations including fisheries.

I would like to thank Department of Science and Technology, Government India, New Delhi for the award of Book Writing Project to me which triggered my interest to pen down my long research and teaching experience for the benefit of those who are interested in this subject. Finally, I acknowledge the support and encouragement of my wife Urmila, daughter Alpana and son Puneet during the progress of this work.

BRIJRAJ KRISHNA DAS
Panjab University
Chandigarh

CONTENTS

Chapter 1

THE HYDROLOGICAL CYCLE

The total volume of water that exists on the earth, approximately 80 percent is in the oceans, 90 percent is in the pores of rocks beneath the earth's surface, 1 percent is in the form of ice, 0.002 percent is in streams and lakes and only about 0.0008 percent is in the atmosphere (Garrels and Mackenzie 1971). Although rivers and lakes make up only a small fraction of the hydrosphere, the rate of water circulation through them is quite rapid. Garrels and Mackenzie calculated that the amount of water discharged from rivers to the sea each year is approximately equal to the total amount of water present in the rivers and lakes. The hydrological cycle illustrates the way the water moves through these various environments (Fig 1.1).

Oceans play a vital role in hydrological cycle as they contain maximum volume of water occupying more than two-thirds of the surface of the earth. In comparison, the biosphere stores only 0.0001 percent of water or 2000 km^3 of all available water (Berner and Berner, 1987). Rivers have only 0.0001 percent of the total water whereas groundwater contributes 1.05 percent of water to the hydrological cycle. Ice caps and glaciers including the polar region, are the largest storehouse of fresh water, contributing 2.97 percent or 43.4 million ($X10^6$) Km^3 to the water budget (Subramanian, 2000).

Water does not remain in any one reservoir, but is continually moving from one place to another in the hydrological cycle. Water is evaporated from the oceans and the land into the atmosphere where it remains for only a short time, on the average about eleven days, before falling back to the surface as snow or rain. Part of the water falling on to the continents runs off in rivers and in some places accumulates temporarily in lakes. Some also passes underground, only to come up later in rivers and lakes. The remaining portion of the precipitation on the continents is returned directly to the atmosphere via evaporation. Over the oceans evaporation exceeds precipitation with the difference being made up by runoff from the continents.

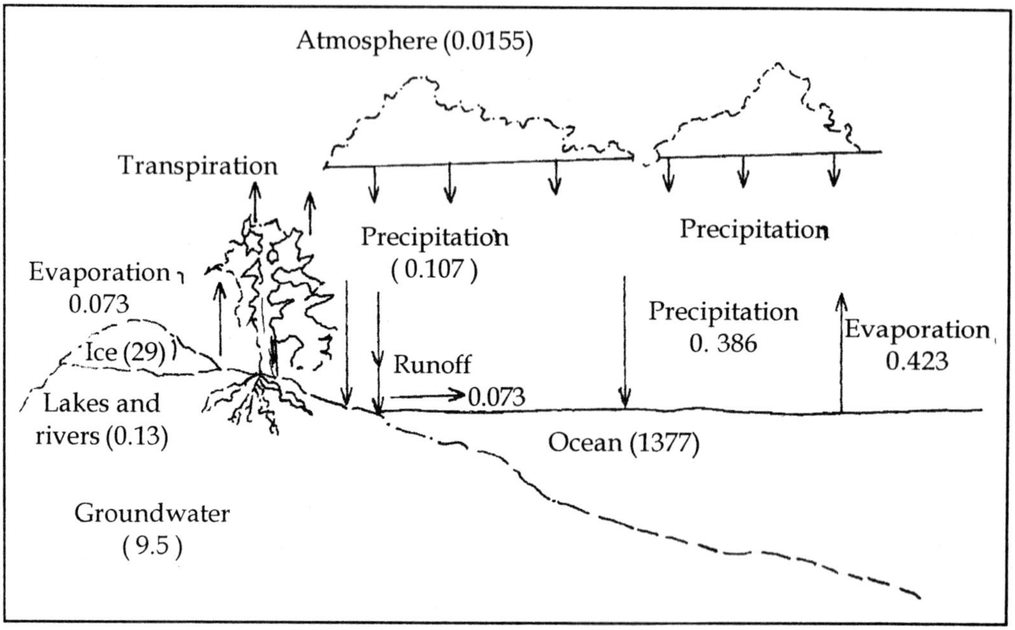

Fig. 1.1. The hydrological cycle. Numbers in parentheses represent inventories (in $10^6 km^3$) for each reservoir. Fluxes are in $10^6 km^3/yr$ (after Berner and Berner, 1987).

In order to conserve total water, evaporation must balance precipitation for the earth as a whole since the total mass of water at the earth's surface is believed to be constant over time. The average global precipitation rate, which is equal to the evaporation rate, is 496,000 km^3/yr. For any one portion of the earth, by contrast, evaporation and precipitation generally do not balance. On the land, or continental part of the earth, the precipitation rate (110, 300 km^3/yr) exceeds the evaporation rate (72,900 km^3yr) whereas over the oceans evaporation 423,100 km^3/yr) dominates over precipitation (385,700 km^3/yr). The difference in each case (374,00 km^3/yr) comprises water transported from the oceans to the continents as atmospheric water vapor, or that returned to the oceans as river runoff and a minor amount of direct groundwater discharge to the oceans.

The residence time is defined as the volume of water in a reservoir divided by the rate of addition (or loss) of water to (from) it. It is the average time a water molecule spends in a reservoir. For the oceans, the volume of water present ($1370 x 10^6 km^3$) divided by the rate of river runoff to the oceans (0.037

x 10^6 km^3/yr) gives residence time of 37,000 years (Berner & Berner, 1987) (Figs. 1.2 & 1.3). This long residence time which may be considered as a filling or replacement time, reflects the very large volume of water in the oceans. By contrast, the residence time of water in the atmosphere relative to evaporation from both the oceans and the continents, is only ten days. Lakes, rivers, glaciers, and shallow groundwater have residence times lying between these two extremes, but because of extreme variability, the average residence time for each of these reservoirs cannot be estimated.

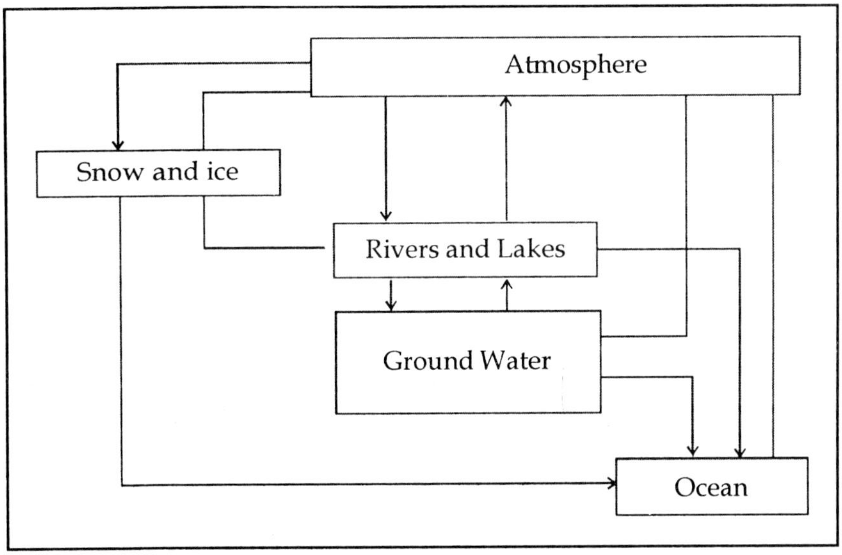

Fig. 1.2. Schematic diagram for the hydrological cycle (taken from Drever, 1982).

HYDROLOGY

The rain that falls on land has various functions. A very small portion of water will remain as droplets on vegetation or remain sprinkled close to the top of the soil and evaporate soon after rainfall stops. A part of it will be taken up by the roots of plants and evaporated through leaves, a process termed as transpiration. The term evapotranspiration is used for the combined processes of evaporation and transpiration. Some of the water may infiltrate into the soil, where it migrates laterally towards a stream, a process called underflow, or it may percolate downward to the groundwater system. When a rainfall is too heavy or prolonged for infiltration to accommodate the water, water will flow on the surface of the ground as overland flow.

The chemistry of rainwater changes drastically as it comes in contact with the earth's surface. Even when it is in the atmosphere before it reaches the ground, rain may pick up solutes from plants, and much of its acidity may be neutralized. As water passes through the soil zone it acquires solutes from dissolution or partial dissolution of minerals, and some solutes (notably nitrogen compounds and phosphate) are extracted from the water by plants. The organisms of the soil release soluble organic compounds to the water, and these compounds accelerate the breakdown of minerals. During the absence of rainfall, evapotranspiration removes essentially pure water from the soil, so that any solutes in the water, either brought in by rain or dissolved from soil minerals, may get concentrated and even precipitate as solid phases.

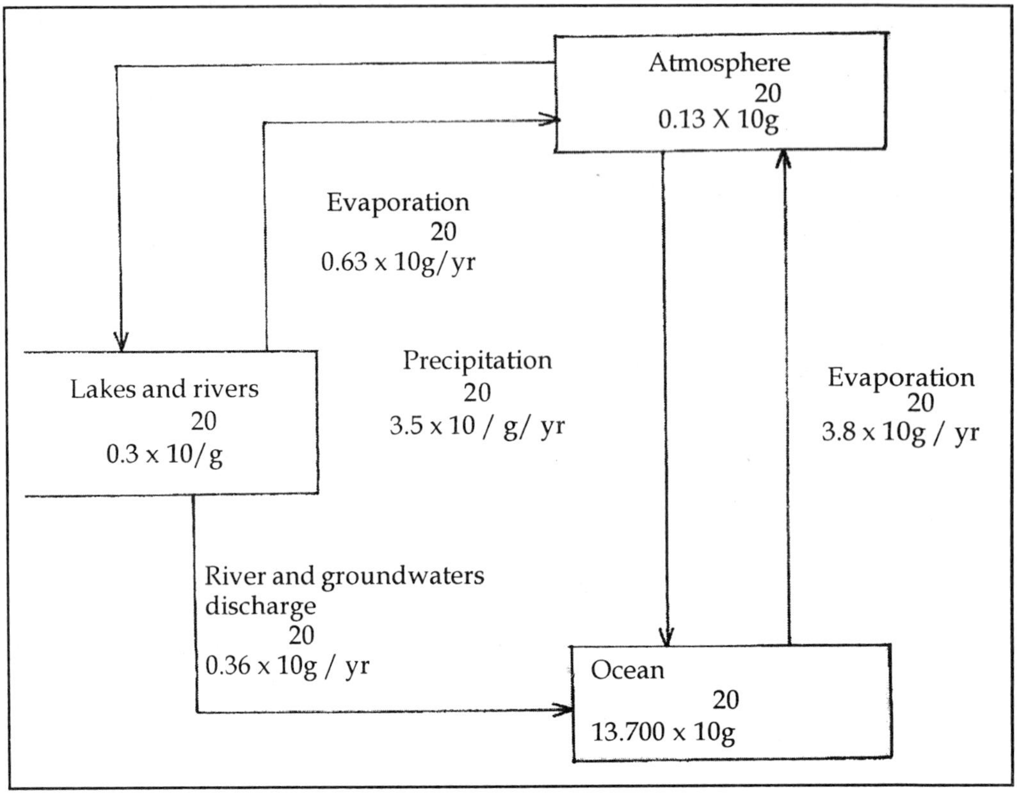

Fig. 1.3. Outline of a more complex box model of the hydrological cycle. (Numbers are from Garrels and Mackenzie, 1971) (after Drever, 1982).

The water in streams and rivers in derived from several sources. During the absence of rain, the water in the stream (it base flow) is derived from the permanent groundwater system. Immediately, after rainfall, the base flow will be augmented by contributions from interflow, overland flow, and of course, rain falling directly on the stream channel. As a stream rises, water often overflows from the channel into porous alluvium that forms its banks, a process termed bank storage. As the stream recedes, water flows back from bank storage into the channel. When the total amount of water leaves an area in streams or rivers, the term runoff is generally used. The ratio of runoff to precipitation varies greatly from region to region depending on climate, geology, and vegetation (Drever, 1988).

GROUNDWATER

The water falling on the ground percolating downward through the soil eventually encounters a level in the soil or underlying bed rock where all pore space is filled with water. At this point water become groundwater, the rock or soil is said to be saturated with water, and the level where this occurs is known as the water table. The water table is the surface that divides rocks in which the pores are completely filled with water from those in which the pores are partially filled with air, the latter is known as under saturated zone or zone of aeration (Fig. 1.4).

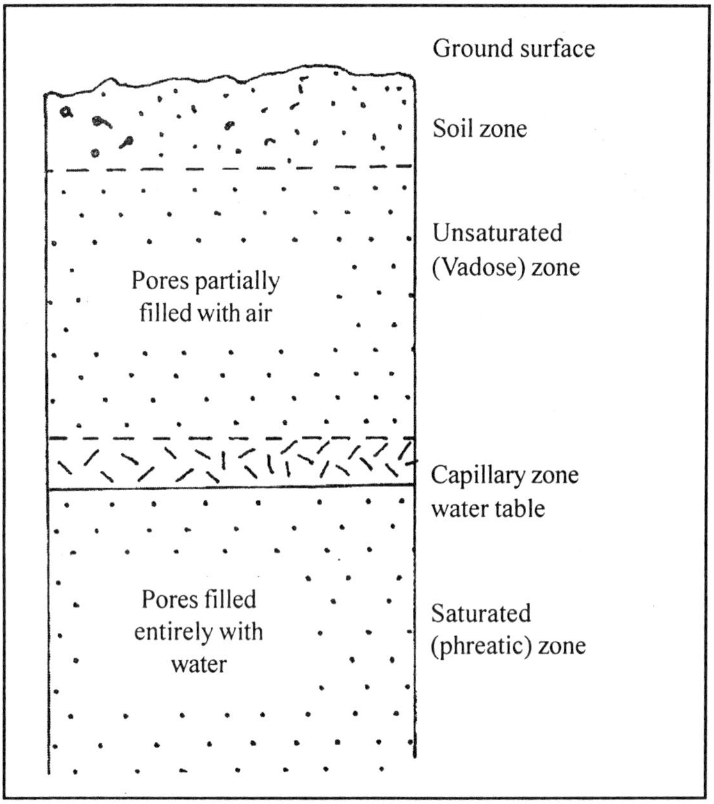

Fig. 1.4. Represent subsurface features in permeable rock (taken from Drever, 1982).

Its position conforms roughly to the configuration of the topography i.e. it is close to the surface in depressions and valleys and deeper on hilltops and prominences. The depth of water table depends on many factors. In areas with moderate rainfall and gentle slope it may be within a few meters of the surface; in rain deficient regions as deep as 50 meter; in the forested coarse-textured soil, quite near the surface, and deep if the soil texture is fine and without vegetal cover. The water table fluctuates conspicuously from season to season, and situation to situation. It rises after abundant rainfall, sinks during dry summers or during drought, and after heavy withdrawal of groundwater through pumping. The region below the water table is called phreatic or saturated zone and region above the water table is the vadose or unsaturated zone. The behavior of groundwater systems is generally controlled by the porosity and permeability of the rocks involved. Porosity is the fraction (or percent) of the total rock that is void space; permeability is a measure of the ability of a rock to transmit fluids. A coarse, uncemented sandstone will typically have a high porosity and high permeability because the pores are larger and interconnected. A shale may have a high porosity, but its permeability is likely to be small because the pores are so small that water, can flow through them only very slowly. Limestones often have quite low porosities but very high permeabilities because the pores are large solution channels or fractures. Even igneous rocks such as granite may have significant permeability resulting from fractures and joints in the rock. An aquifer is a water saturated rock with sufficient porosity and permeability to be a usable source of water for wells. There are three types of aquifers: unconfined or water table aquifer, confined aquifer and artesian or pressure aquifers. Unconfined aquifer is one where the water table is not overlain by impervious soil and not directly exposed to the atmosphere. When water is trapped between two impermeable layers it is called confined aquifer where it is under very high pressure. A perched water table may occur where saturated permeable rocks are separated from unsaturated rocks below by an impermeable barrier (Fig. 1.5).

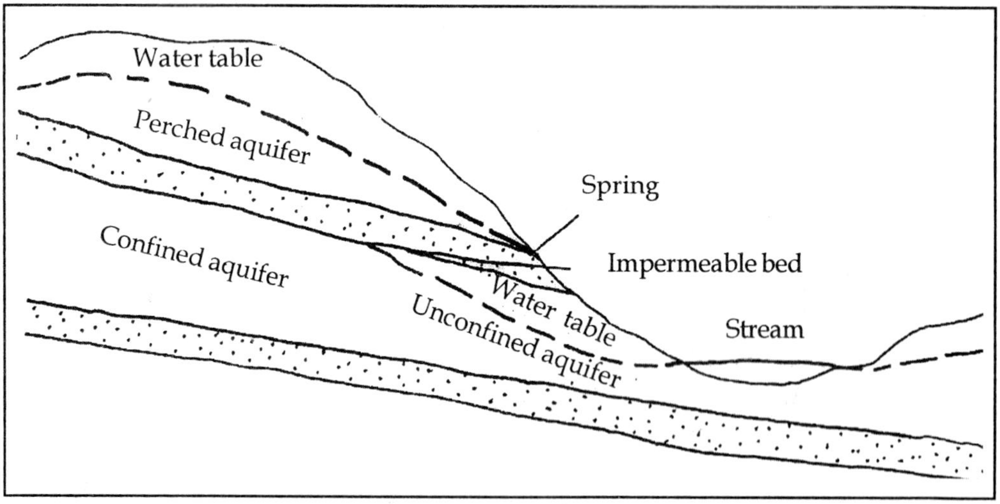

Fig. 1.5. Perched and confined aquifers (after Drever, 1982).

If a well is drilled into a confined aquifer the water in the well will rise to a level above the top of the aquifer. The level to which the water would rise in series of such wells is the piezometric surface. If the piezometric surface is above ground level (Fig.1.6), water will flow from the well without pumping. If the discharge from wells is greater than the natural recharge, the water in the aquifer will be depleted and the piezometric surface will be progressively lowered.

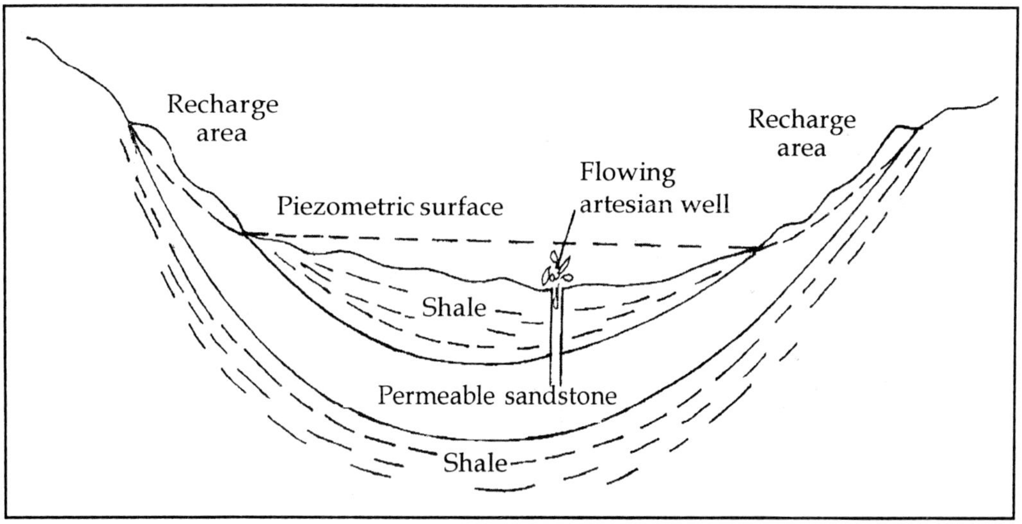

Fig. 1.6. Diagramatic representation of a artesian system in a syncline. The piezometric surface will become depressed if significant amounts of water are withdrawn from the aquifer (after Drever, 1982).

Chapter **2**

LAKE TYPES

The lakes are formed by various natural processes besides those, which are man made. The latter type of lakes are generally shallow and are fed from rainwater or manual filling and get almost dried up during summer season. If there are no outlets, the water becomes dirty and polluted. It has been noticed that the lakes formed by tectonic activity are the deepest lakes and contain enormous amount of water. Therefore, the lakes have their own characteristic depending upon their origin.

TECTONIC LAKES

The largest (deepest) lakes are those of tectonic or structural origin. The major type of tectonic lakes form as a result of faulting in which depressions occur between the masses of a single fault, displacement or in down faulted troughs by normal faulting and block faulting. The latter type of basin is referred to as a graben, and is the mode of origin of a large number of the most spectacular relief lakes of the world (Fig.2.1).

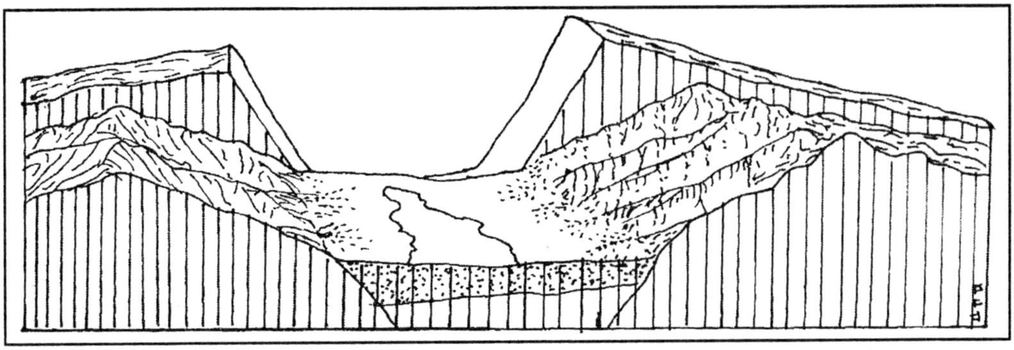

Fig. 2.1. Diagramatic representation of faulted trough graben leading to lake formation.

Foremost among these is lake Baikal of eastern Siberia formed in the Baikal Rift of the Siberian Platform, the deepest lake in the world. Several other lakes, such as Tanganyika, Lake Malawi and Lake Mobutu are formed along the East African Rift Valley.

Several lakes in Himalaya are of tectonic origin. In Kumaun, Lesser Himalaya, Lake Nainital formed due to neotectonic activity in Holocene period (10,000 years ago). The movement along the Nainital fault caused upliftment of northern block relative to the southern and the vertical movement was rotational, leading to blocking of a perennial stream (Balia river) and gave birth to Nainital lake (Fig.2.2).

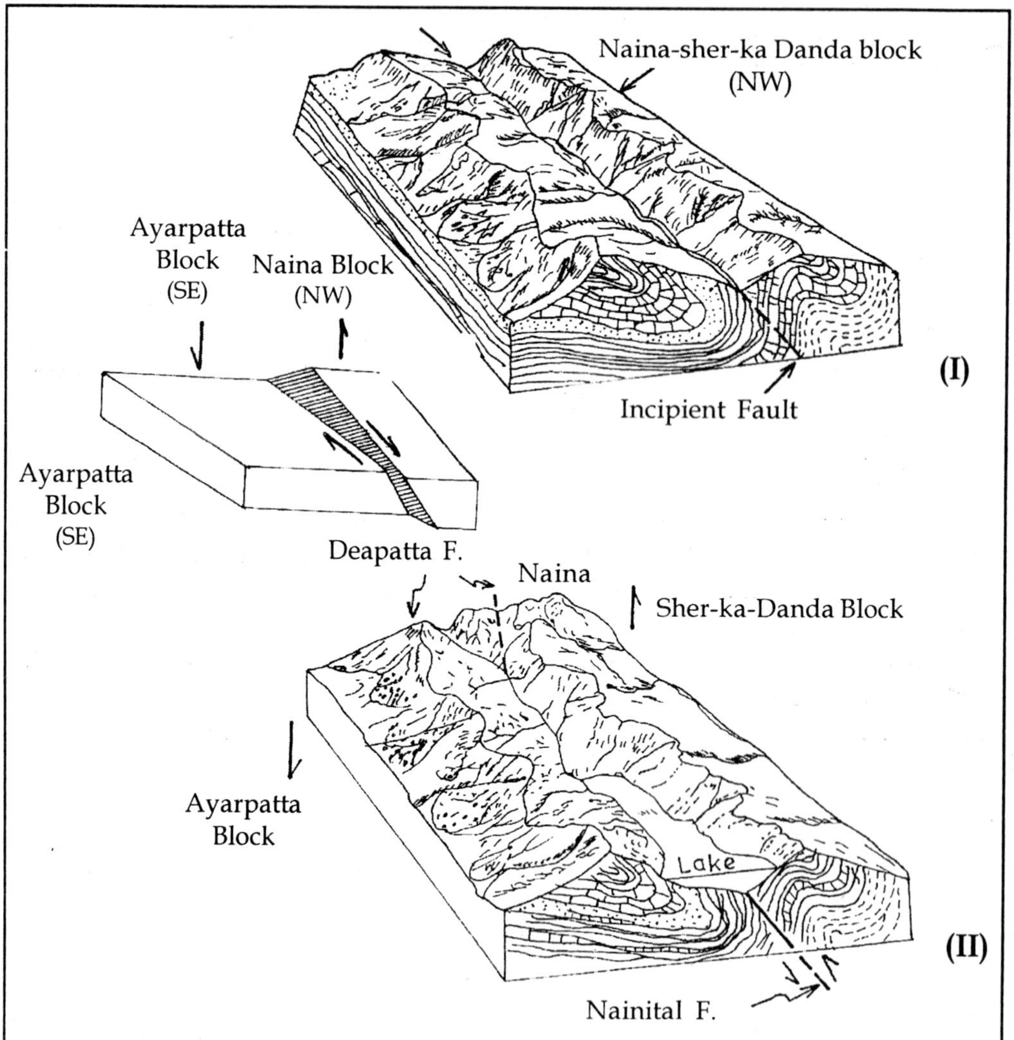

Fig. 2.2. Digramatic representation of orgin of Nainital Lake. The tilting of a uplifted Sher-Ka-Danda-Naina ridge blocked the flow of a perenial Balia river (after Valdiya, 1988).

Similarly, Bhimtal and Naukuchiyatal lakes of Kumaun formed due to tectonic movement along a WNW-ESE trending fault (Main Boundary Thrust) in the South and Ramgarh Thrust in the north, which caused the southern block to sink relative to north giving rise to depression. Parallel to the Bhimtal Fault there is another transverse fault along Sattal which caused landmass to the east of the fault to subside, resulting in the damming up of the stream and the development of depression and hence the Sattal Lake is formed (Raina and Dungrakoti 1975; Valdiya, 1980, 1988) (Fig.2.3).

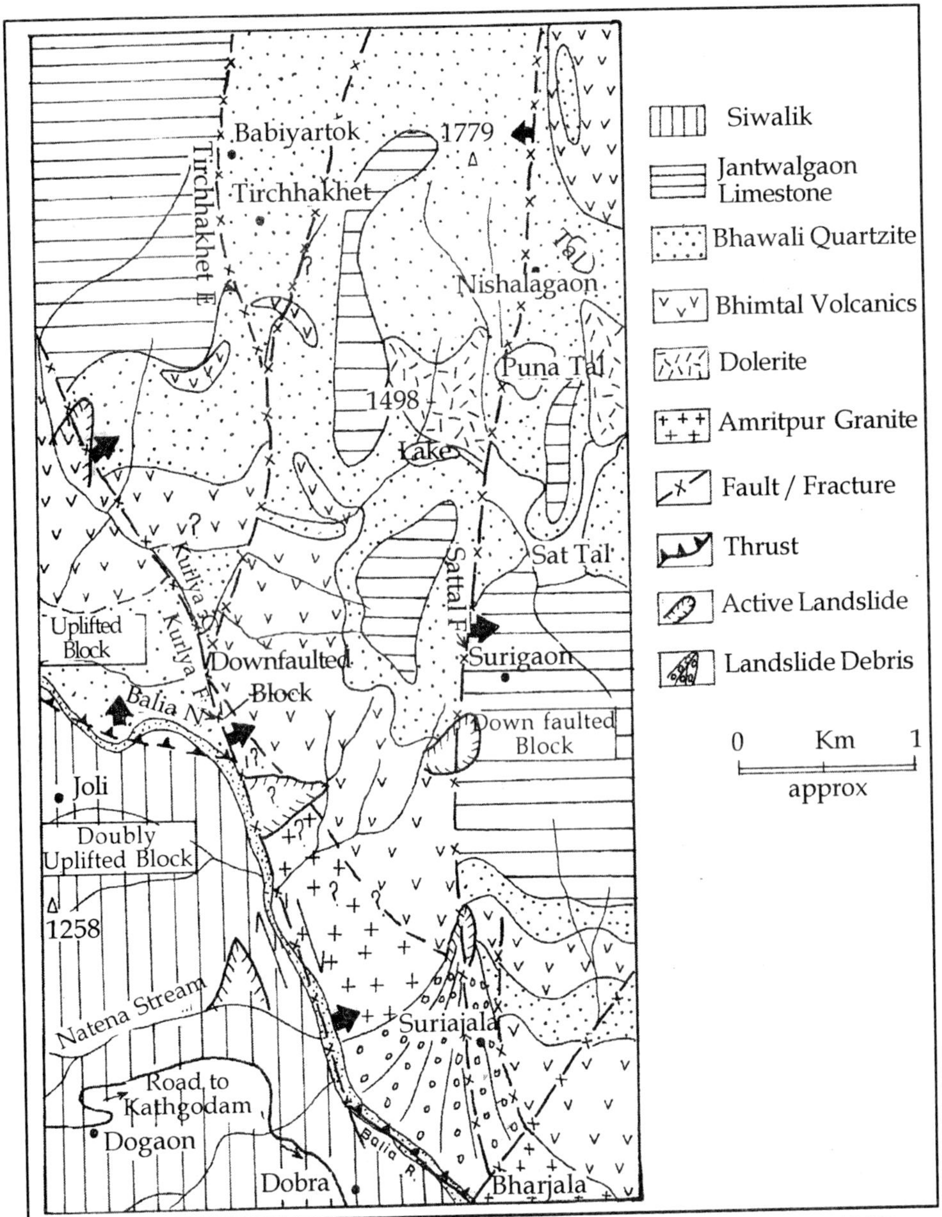

Fig. 2.3. Map Showing tectonic origin of Sattal Lake (after Valdiya, 1988).

The Mansar and Surinsar lakes of Jammu also came into existence due to damming up of the river flowing along the strike valley and a fault running is the same strike trend blocking the river resulting in the formation of lakes. These lakes were once probably continuous and got separated by the neotectonic uplift during Holocene (Zutshi, 1985). Renuka lake in Sirmaur district, Himachal Pradesh, is of tectonic origin, formed along the abandoned course of Giri River due to tectonic upliftment. It developed in a tectonic valley as a consequence of faults, Krol Thrust and Giri Thrust and is fed by the tributaries of the Yamuna and the Giri rivers (Fig.2.4). Lake Nilang in Kashmir at an altitude of 2180 meters is believed to have formed due to the recent tectonic activity (Khan and Zutshi,1980).

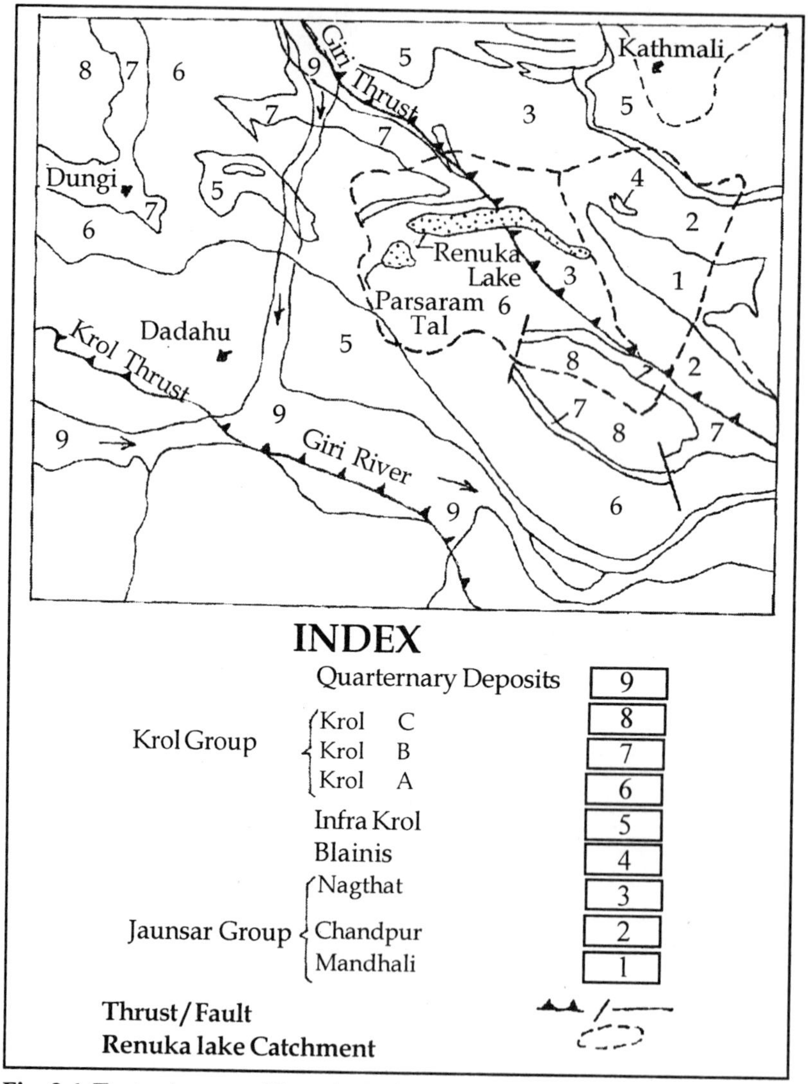

Fig. 2.4. Tectonic map of Renuka Lake representing its origin, lithology and catchment area.

Lakes Formed by Volcanic Activity

As volcanic materials are ejected upward and create a void, or as released magma cools and is distorted in various ways, depressions and cavities are created. If these depressions are undrained they may contain a lake. Small crater lakes are occasionally found occupying cinder cones of quiescent volcanic peaks. However, crater lakes in depressions formed by the violent ejection of magma, or by collapse of overlying materials where underlying magma has been ejected to create cavity, are more common in all areas of recent volcanic activity. The Lonar crater lake of Maharashtra State (India), is a circular depression 135 meter deep, 1200 meter across lying amidst sub-horizontal volcanic rocks of the Deccan Trap age. The lake is supposed to be late Pliocene to mid-Pliocene in age (La Fond and Dietz, 1964) and have formed by explosive impact of a meteorite (Beals, et al., 1960; Arogya Swamy, 1962; Venkatesh, 1965, 1967; Nayak, 1972; Fredrickson et al., 1973). Some volcanic lakes formed by a combination of large scale volcanic and tectonic processes. In some situations Caldera Collapse occurred on such a large scale that extensive portions of the surrounding land subsides in addition to the central portion of the volcano. Such subsidence usually takes place along preexisting fault fractures. Some of the larger lakes associated with volcanic activity were formed in this manner (Witzel, 1983) (Fig.2.5).

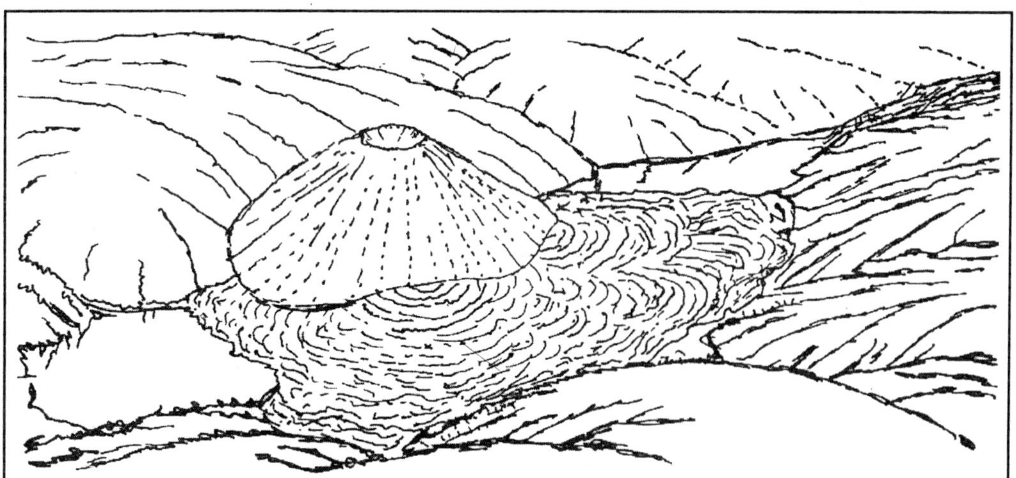

Fig. 2.5. Diagramatic representation of crator volcanic lake.

Lakes Formed by Landslides

Sudden movements of large quantities of unconsolidated material in the form of landslides into the floors of stream valley can cause dams and create lakes, often of very large size. The landslides usually are brought about by excessive rains acting on unstable slopes and also initiated by earthquake activity. Such lakes, however, are transitory, for the unconsolidated material of the slide may be easily eroded by the effluent of the lake once the latter rises to spill over the top of the dam (Hutchinson, 1957). A small Gohana Tal lake formed due to landslide which blocked the flow of a tributary of the Alaknanda river system in Chamoli district of North-Eastern Garhwal. Heavy rains in July 1970 caused the natural dam to burst resulting in devastating flash floods in the entire Alaknanda Valley.

Lakes Formed by Glacial Activity

With retreat of the last Pleistocene glaciers an immense number of small lakes were created. Lakes of glacial origin are far more numerous than lakes formed by other processes. The action of glaciers in mountainous regions of high relief usually produces lake basins that are quite different from those that result from movements of large ice sheets on regions of more mature and gentle relief. Lakes formed as a result of glaciation may have been produced in a number of ways by ice damming, by drainage barriers formed of moraine materials; by ice scour; by freeze thaw effects at the head of glaciated valleys (Cirque Lakes), and by valley glaciation which produced long, narrow, and deep lakes such as fiord lakes of many high latitude regions of both northern and southern hemispheres.

Lakes situated between 3000 and 4000 meters in Himalaya are glacial which originated during the third Himalayan glaciation. The high mountain lakes are generally small and shallow situated above the tree line and fed by glaciers. The summer monsoon is weak and the water in lakes is primarily maintained by snow melt and local precipitation (Zutshi, 1991).

A number of glacial lakes are known from the Higher Himalayas. Bhagsunath Lake located on the upper reaches of Dhauladhar range in the Dharmshala area of Himachal Pradesh is fed by snow melt waters from the surrounding areas. Chandra Nahan Lake lies at the origin of the Pabar river which is a tributary of the Tons River to the north-west of Rhoru in Himachal Pradesh. It lies in a depression formed by the glacier that feeds the Pabar river and the Lake is fed by snow-melt waters. Another beautiful glacial lake is located at the source of Chandra river in the Lahul Valley of Himachal Pradesh. It lies in a large depression formed by the glacier and is fed by melt waters from Chandra glacier. The lake freezes during the winter season.

In Garhwal, Deori Tal Lake located in Chamoli district, is a depression formed by glacial movement in the past. It is fed by a snow-melt stream. Hemkund-Lokpal lake is a famous lake, due to Sikh pilgrimage, located in the upper part of the Chamoli district in Garhwal. It is a large glacial lake lying on the cirque floor of the Saptsring glacier and fed by a stream from a hanging glacier located on the eastern slope of the Saptsring peak. Another similar lake is Kagbhusand Tal lake located on the slope of the Kankulkhal peak in the north-eastern Chamoli in Garhwal. It occupies a crique of the Kankulkhal glacier which also acts as its feeder. Another famous glacial lake is Manimahesh located at elevation of 4200 meter in the Budhil Valley near Bharmour in Chamba District, Himachal Pradesh. The lake lies in a glacial depression and is fed by snow-melt water descending from the surrounding slopes. Tsomorari lake is located in the Rupshu area of Ladakh between the Zanskar range in south and the Ladakh range in the north. It lies in a vast glacial depression and is fed from the snow-melt waters flowing into it from different directions. It has no outlet and hence the water is brackish.

Lake Basins Formed by River Activity

The running waters of river possess considerable erosive power that may create lake basins along their courses from elevated land to large lakes. A combination of destructive erosional and obstructive depositional processes occurs as rivers flow down more gentle slopes to form many lakes in the river flood plains. Many lakes were formed along large rivers when sediments of the main stream were deposited as levees across the mouths of tributary streams. Lakes that are formed in the deltas of all major rivers of the world are common.

As a river meanders within the irregularities of the topography, greater turbulence on the inner, convex side where currents and turbulence are reduced with time, continued erosion and concavity takes place until the V-shape meander of the river closes in upon itself. The main course of the river cuts a channel through the initial portion of the meander, and levee deposits eventually isolate the loope, referred to as an oxbow lake from the river channel (Fig.2.6).

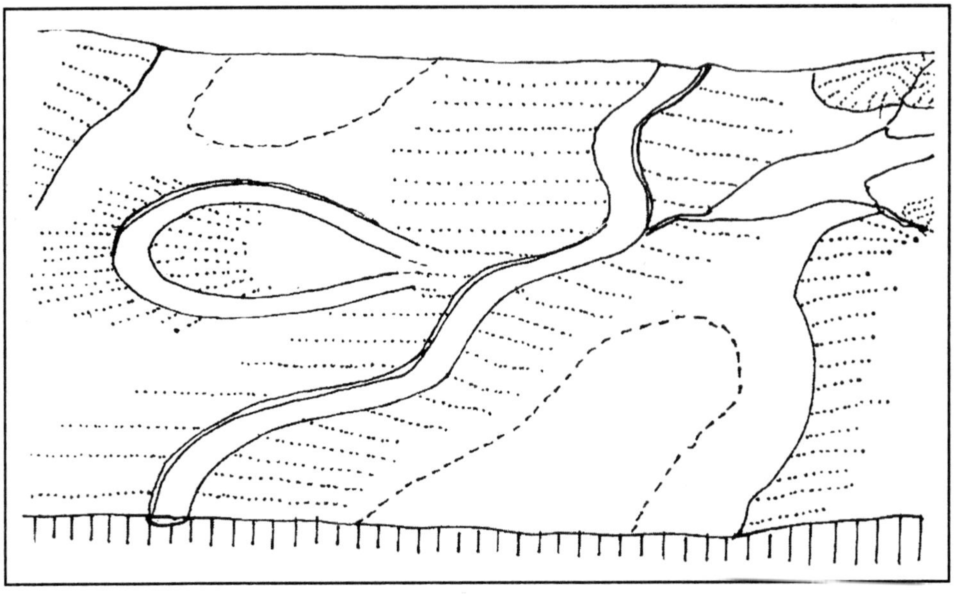

Fig. 2.6. Sketch Showing mechanism of formation of Oxbow Lake.

The outer erosional side of oxbow lakes is usually deeper than the inner concave side. The Manasbal and Naranbagh lakes of Kashmir valley situated within the flood plan of the River Jhelum at an altitude of 1584 meter are supposed to have originated from the meandering of alluvial materials (Khan and Zutshi, 1992; Zutshi et al, 1980). These are mostly shallow and differ appreciably in surface area and depth.

□□□

Chapter 3

PHYSICAL PROCESSES IN LAKES

A brief outline of some fundamental aspects of lakes and the processes, of physical, biological, geological and anthropogenic which affects the lakes, because of their generally small size, are discussed below to understand dynamics of lake system.

WATER BALANCE

Lakes are closed or open. Those lakes in which water leaves only by evaporation are called closed and those which has outlets, open. Some lakes which do not have exposed outlets and water passes out as superficial ground water are known as seepage lake in contrast to drainage lake which receive water from rivers and streams (Hutchinson, 1957).

Lakes have different processes of inputs and outputs of water into and from the lakes. In arid regions lakes often have no outlet and water is lost only by evaporation. Inputs can also vary considerably. Lake Mansar in Jammu, India, is deep and extensive lake, located in Lesser Himalaya with moderate to high rainfall. The lake receives water from rainfall and underground source. The water level of lake changes very little compared to its volume and depth during summer season, except for minor fluctuations. Lake Victoria in Africa is a shallow but extensive lake with minor stream input is located in the region of high rainfall. Nearly three-fourths of the water input to the lake is provided by rainfall on the lake surface (Hutchinson, 1957). Commonly for most larger lakes, major inputs are from rivers and streams with rainfall contributing for only a small percent of the total. In contrast to this, some small lakes like Khajjiar in Chamba district and some Higher

Himalayan lakes in Spiti valley are almost spring-fed and water is totally supplied by underground springs and /or snow melt in case of latter. In limestone terrain, Karst lakes develop because of the permeable nature of underlying partly dissolved bed rock (limestone) due to which water is able to readily seep into and out of the lake floor. However, in many lakes because of the presence of relatively impermeable clay-sediment on their bottom underground, seepage does not occur.

The volume of water in a lake, and consequently the level of its surface, depends on how the inputs and outputs of water balance. If there is a good balance, the water level is maintained and fluctuates very little. However, because of year to year variation in rainfall and temperature (climatic factors), some fluctuation takes place. The degree to which the fluctuation affects the lake level depends on the size of the lake and the rate of input and output of water.

If a lake maintains a constant volume, because inputs and outputs are equal, it is said to be in a steady state with respect to its water content. In this case the replacement time can also be viewed as a residence time of water in the lake. This is the average time a water molecule spends in the lake before being removed through the outlet (Berner and Berner 1987).

Thermal Regimes and Lake Classification

The chemistry of fresh water lakes depends to a large extent upon a circulation driven by temperature changes. The density of water in lakes is primarily a function of temperature (Fig.3.1). As temperature changes, density changes, and if less dense water becomes overlain by more dense water, convection, or lake water overturn, takes place (See chapter 5, Mansar Lake). With heating, instead of decreasing in density like most liquids, dilute (fresh) water actually increases in density over this temperature range. At 4^0C, the water is at a maximum density and above this temperature, the density then continues to decrease with temperature like a normal liquid. This unusual temperature-density behavior of fresh water provides the fundamental basis for classifying lakes and for describing their overall circulation behavior. (Berner and Berner, 1987).

The upper layer or epilimnion is warm as a result of solar radiation. The temperature within the epilimnion is fairly uniform because the surface zone is stirred by wave action. Below the epilimnion we encounter a region where temperature falls rather rapidly with depth. This is known as thermocline or mesolimnion. Below the thermocline is the deep, cold portion of the lake known as the hypolimnion (Fig.3.2). In reality, the situation may be complicated by the presence of more than one metalimnion, and, of course, very shallow lakes and ponds will consist of an epilimnion only. The situation shown in Fig.3.2 represents stable stratification. Warm, low-density water floats above cold, high-density water, and the density contrast results in little mixing or exchange of solutes between epilimnion and hypolimnion.

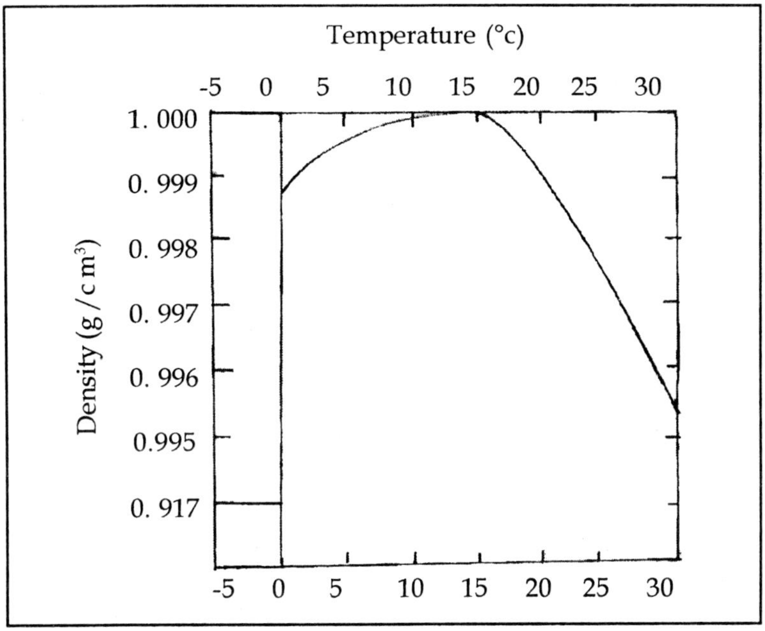

Fig. 3.1. Density of pure water at 1 atmosphere showing function of temperature (after Berner and Berner, 1987).

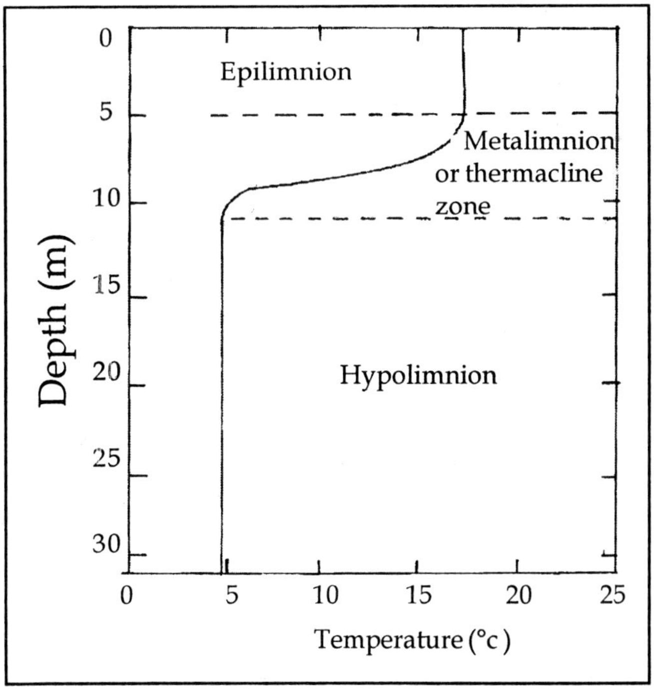

Fig. 3.2. Temperature profile of a typical temperate freshwater lake in summer (taken from Berner and Berner, 1987).

At the end of summer, the temperature of the epilimnion starts to decrease. When it reaches a value close to that of hypolimnion (about 4 °C in deep lakes in cold-winter climates), the density contrast between epilimnion and hypolimnion disappears. The winds cause complete mixing between the epilimnion and hypolimnion and this event is called the fall turnover. If surface temperatures become even colder, stable stratification may again develop, with water at about 0°C overlying more dense water at 4°C. This stratification disappears at the spring turnover, when the surface layers again warm at 4°C.

At times of turnover, the dissolved oxygen content of the entire lake is essentially determined by equilibrium with oxygen in the atmosphere. When the lake becomes stratified, the oxygen content of the hypolimnion steadily decreases as a consequence of decomposition of organic matter falling from the epilimnion. The amount of organic matter produced in the epilimnion is largely controlled by the availability of inorganic nutrients particularly phosphate. A lake in which the supply of nutrients (nitrogen & phosphorus) is low, due to which there is little photosynthetic production and the water is oxygen-rich at all depths is known as, oligotrophic lake. Such lakes are usually deep and have relatively plankton-free clear water. In contrast to this, a eutrophic lake is one in which the supply of nutrients is high, photosynthetic activity is high and the hypolimnion is anaerobic. Eutrophic lakes have high concentrations of plant nutrients and large concentrations of plankton due to high organic productivity. In an oligotrophic lake the pe (redox intensity) of the hypolimnion remains high; in a eutrophic lake it will decrease with depth and with time since the onset of the stratification. When turnover occurs in a eutrophic lake, hydrogen sulfide and other toxic compounds may be mixed into the epilimnion, which may cause wide spread fish mortality in the epilimnion.

When an oligotrophic lake becomes eutrophic as a result of pollutions, it is difficult to reverse the process. This is because of reduction of ferric compounds. Ferric oxides absorb phosphate strongly, and when the oxides are reduced (to Fe^{2+}, FeS_2 or $FeCO_3$), a large amount of phosphate is released to solution (Williams *et al.*, 1976). Thus, when a lake first becomes eutrophic, phosphate is transferred from the sediment to the hypolimnion, and when turnover occurs, this additional phosphate will be mixed into the epilimnion. The extra supply of phosphate will further aggravate the problem of eutrophication.

In many tropical lakes, temperature differences between night and day are sufficient to cause frequent mixing, and hence prevent stratification. Thermal stratification may occur when a mass of cold water is formed during an unusually cold period, and a permanent epilimnion forms above it.

Photosynthesis, Respiration, and Biological cycling

Photosynthesis is the process by which carbon dioxide (which may be derived from HCO^-_3) is converted to organic matter and oxygen:

$$CO_2 \xrightarrow{\text{sunlight}} C_{\text{organic}} + O_2$$

Photosynthesis used energy from the sun to form compounds (organic matter and oxygen) that are unstable together from the stable compound, CO_2. In the absence of photosynthesis, the oxygen in the atmosphere would be steadily consumed by the oxidation of organic matter until no free oxygen remained (Garrels and Perry, 1974; Holland, 1978).

Carbon and oxygen are not the only elements involved in photosynthesis. Plants also require nitrogen compounds, phosphorus compounds, and a wide range of trace elements. Plants (in aquatic systems primarily, microscopic algae), which grow and multiply until all available phosphate or nitrate is used up, at which point photosynthesis ceases. In unpolluted fresh water, phosphate is normally the limiting nutrient; nitrate is some times limiting in polluted waters. The availability of particular trace elements is occasionally limiting for short periods of time, but this is rare. Photosynthesis is the ultimate source of food for all organisms in surface waters. The overall process of photosynthesis in lakes (and the ocean) can be represented by the following reaction (e.g. Stumm and Morgan, 1981):

$$106\ CO_2 + 16NO^-_3 + HPO^-_4 + 122\ H_2O + 18\ H^+ = \text{(+ trace elements and energy)}\ C106\ H_{263}\ O_{110}\ N_{16}\ P_1 + 138\ O_2\text{algae}$$

This equation illustrates the enormous importance of phosphate in the carbon-oxygen balance of waters, particularly lakes. Each atom of phosphorus (as phosphate) added to a surface water results in fixation of about 106 atoms of carbon in organic matter, and when the organic matter produced from one atom of phosphorus decays, it has the potential to consume 138 molecules of oxygen.

Along with photosynthesis, the process of aerobic respiration also takes place. Respiration occurs in all organisms including plants, animals, and bacteria, and, in the presence of O_2, it is the reverse of photosynthesis:

$$C106\ H263\ O110\ N16P + 138\ O_2 = 106\ CO_2 + 16NO^-_3 + HPO_4^{2-} + 122\ H_2O + 18\ H^+$$

and involves the breakdown of organic matter while photosynthesis involves its formation. Photosynthesis can occur only if water is sufficiently shallow, so that light, necessary for the process, can penetrate it. At greater depths it is absent. In sufficiently deep (or murky) lakes, respiration dominates over photosynthesis at depth and this helps to destroy much of the dead organic matter sedimenting from above. In other words, in sufficiently deep lakes the processes of photosynthesis and respiration are in part separated from one another with net photosynthesis at the surface and net respiration at depth.

Carbon is released as CO_2, organically combined nitrogen as NO^-_3, and organically combined phosphorus as HPO_4^{2-}. The release of CO_2 causes an increase in PCO_2 (partial pressure of CO_2) and hence a decrease in pH.

When molecular oxygen is not available, or when it has been used up, decay of organic matter continues by a series of reactions that represent successively lower pe (redox intensity) levels. Some common reactions takes place as follows:

Denitrification

In denitrification, bacteria use the oxygen in nitrate ion to oxidize organic carbon to CO_2. The stoichiometry of the reaction is 2.5 C organic + $2NO_3^-$ $+2NO_3^- + 2H^+ = N_2 + 2.5\ CO_2 + H_2O$

The importance of this reaction is that it converts nitrate, an essential nutrient, into biologically inert molecular nitrogen, which is important in the nutrient balance of lakes and rivers.

Deamination of Amino Acids

Amino acids are the constituents of protein, and in anaerobic conditions, they decompose according to the stoichiometry

$$[\text{Amino acid}] = [\text{N - free compounds}] + NH_3$$

The NH_3 reacts to from NH^+_4,

$$NH_3 + H_2O + CO_2 = NH^+_4 + HCO^-_3$$

causing a net rise in pH. The rise in pH may cause precipitation of calcium as a carbonate or as a soap (the calcium salt of a fatty acid) (Berner 1968, 1971), which may be important is formation of concretions.

Sulphate Reduction

Sulphate reducation is a bacterial reaction in which bacteria use the oxygen in SO_4^{2-} to oxidize organic matter to CO_2, producing sulphide species as a by-product. The stoichiometry of reaction is $SO_4^{2-} + 2C_{\text{Organic}} + 2H_2O = H_2S + 2HCO_3^-$.$HS^-$ will form rather than H_2S if the pH is above 7, and if any reactive iron compounds are present, the sulphide species will react with them to form solid sulphides.

Fermentation Reactions

There are many bacterial fermentation reactions of the general type which can occur at various pe values, depending on the particular organic compounds involved. The simplest (and one of the most important) fermentation reaction is methane generation,

$$2C_{Organic} + 2H_2O = CO_2 + CH_4$$

The various anaerobic decay reactions are all medicated by bacteria, which use the reactions as a source of energy. Since the bacteria derive energy from the reactions, their function is essentially to catalyze the conversion of thermodynamically unstable systems to more stable systems. Sulphate reduction does not occur until all molecular oxygen has been used up, and methane generation does not occur until all sulphate has been used up.

Eutrophication

The lakes have been classifed as oligotrophic and eutrophic based on concentration of nutrient element such as nitrogen and phosphorus in them. The lake of nutrients results in few plants and, thus, a low rate of organic matter production by photosynthesis. Such lakes, known as oligotrophic lakes are usually deep and have relatively plankton-free clear water, which is well oxygenated at depth (Table 3.1). The eutrophic lakes are well-fed lakes and have high concentrations of plankton due to high organic productivity. Their waters are murky with suspended plankton and often depleted in oxygen at depth. Such lakes depending upon the extent of eutrophication reduces the volume of water and the size of the lake ultimately leading to a marshy land or dryness. Eutrophic lakes are threat to human life as well as aquatic life due to pollution of water and deficiency of oxygen. The nutrients are generally supplied to the lakes from anthropogenic sources such as discharges of effluents from the industries, sewage discharge, throwing of trashes etc. by humans. The Dal lake of Kashmir is a well known example which has converted into eutrophic lake due to various human activities particularly, the cottage industries along the bank and Shikara (boats) in the lake as mentioned above. The Sanasar lake at Patnitop, on way to Kashmir, has completely dried up. The Nainital lake in Kumaun Himalaya is in advance stage of euthrophication because of dense population in the valley and discharge of pollutants by the human beings. Khajjiar lake in Dalhousie, Himachal Pradesh, is also in advance stage of eutrophication by human activity. The lakes which have properties between those that are termed as eutrophic or oligotrophic is termed mesotrophic.

Table 3.1. Criteria of Oligotrophic and Eutrophic Lakes

	Criteria	Oligotrophic	Eutrophic
1	Concs. of plant nutrients (N, P)	"Poorly fed"; low concs. of plant nutrients; P< 10 μg/l	"Well fed"; high concentration of plant nutrients; P> 20 μg /l
2	Organic matter enrichment	Low concs. of organic matter	High concs. of organic matter, either authochtonous (produced in lake) or allochthonous (transported from terrestrial source)
3	Biological productivity (organic matter productivity)in gorg c/m^2/yr generally phytoplankton productivity via photosynthesis	Low productivity – few plants (phytoplankton) because low nutrient concs. Primary productivity < 150g Cm^2/yr; conc. Chlorophyll A (phytoplankton) < 3 μg/l	High productivity, large concs. of plant, particularly plankton (green & blue algae) and/or macrophytes (rooted plants): primary productivity > 250gC/m^2/yr; conc. Chlorophyll A (phytoplankton) > 6μg/l
4	Depth (volume) of lake	Deep, large volume (>15-25m depth)	Shallow, small volume (≤ 10-15m depth)
5	Water	Clear, blue	Murky, dark, turbid
6	O_2 in hypolimnion in summer	Well-oxygenated hypolimnion > 50% O_2 saturation after stratification	Oxgen depleted in hypolimnion by high biological oxygen demand to decompose organic matter < 10% O_2 saturation (~ < 1mg/l O_2) after stratification
7	Bottom fauna	Diversified bottom fauna, deep-water fish, char, white fish, trout	Bottom fauna tolerant of low oxygen conditions, no deep water fish: carp
8	Bottom sediment	Sandy , inorganic, low N	Organic rich much high in N

(Taken from Berner and Berner, 1987)

Limiting Nutrients

Organic compounds produced by algal photosynthesis in lakes consists of carbon, hydrogen, oxygen, nitrogen, and phosphorus. Of the nutrient elements needed for photosynthesis, hydrogen, and oxygen are readily available. Carbon is generally available from atmospheric carbon dioxide. The major elements that are not always available are nitrogen and phosphorus. Relatively small amounts of N and P can produce relatively large amounts of organic matter. The total amount of organic matter produced will be determined by the availability of the nutrient elements that is the least abundant; this is the so-called limiting nutrient and in lakes it is usually phosphorus (less commonly nitrogen) that is limiting. On the basis of calculation of the ratio of demand (amount required) by fresh water plants (algae, diatoms rooted plants) to supply from average river water for various nutrients in a world average lake, the ratio of phosphorus is high (Vallentyne, 1974). This shows that phosphorus is often the limiting nutrient in lakes followed by nitrogen. This explains that the addition of these nutrients could lead to greatly expanded biological productivity and eutrophication. Nitrogen deficiencies (low N/P ratios) can be made up by development in the lake of blue-green algae which are capable of fixing N_2 from the atmosphere.

Source of Phosphorus in Lakes

The primary source of phosphorus (and nitrogen) in lakes are direct rainfall and snowfall on the lake and runoff from the surrounding drainage area. In oligotrophic lakes most phosphorus in runoff comes from rock weathering and soil transport. However, in areas influenced by humans there are additional sources of phosphorus, including agricultural runoff (containing phosphorus from fertilizers and animal wastes) and sewage (containing phosphorus from human waste, detergents, and industrial wastes), which are discharged directly into the lake and its inlet tributaries.

The major source of phosphorus is weathering, that is the removal of phosphorus from rocks. However, phosphorus in rocks and sediments is in a relatively insoluble form as the calcium phosphate mineral apatite. Even when released as soluble phosphate (PO_4^{3-}) by weathering, phosphorus is usually quickly tied up in the soil as iron, aluminum and calcium phosphates or by clay minerals to produce insoluble forms not accessible to plants. Because of the relative insolubility of phosphorus it is often a limiting nutrient in biological systems; that is, it is in short supply. Humans have intervened in the phosphorus cycle by deforestation, which increases erosion; by the use of phosphorus, fertilizers; and through the production of industrial wastes, sewage, and detergents. In lakes, the introduction of greatly increased phosphorus has increased productivity and led to eutrophication.

Phosphorus concentrations in rainfall are small (0.01 - 0.04 mg/l total P) and difficult to measure due to contamination. The amount of phosphorus delivered to land in precipitation has been estimated at 1 TgP/yr (Meybeck, 1982) with total P deposition (including dry deposition) being 3.2 TgP/yr (Graham and Duce, 1979).

❑❑❑

Chapter **4**

WEATHERING AND WATER CHEMISTRY GENERAL

The chemistry of most surface water and groundwater is the result of interaction between rain or snowmelt and rocks near the earth's surface. Much of the interaction takes place in the soil zone. Soil formation is an exceedingly complex process involving plants, animals, microorganisms, and inorganic processes.

The type of soil formed under a particular climate is strongly influenced by the nature of the bedrock, topography, vegetation, and the time over which soil forming processes have operated.

Geochemical approaches to understanding the chemistry of surface waters have generally ignored the complexities of the soil zone. The rain water that has been intercepted by vegetation and then drips off is termed through fall. Water infiltrating the soil is called soil water, and that passing directly into the nearest stream is referred to as surface runoff. Once in the soil, the water either passes downward or is taken up by plant and tree roots. In the latter case, the water is transported up through the tree and eventually evaporated from leaf surfaces. In this way the water is returned to the atmosphere and the overall process is known as transpiration. Water trickling downward through the soil eventually encounters a level in the soil or underlying bedrock where all pore space is filled with water. At this point the water becomes groundwater, the rock or soil is said to be saturated with water, and the level where this occurs is known as the water table (Fig.4.1). Above the water table, pore space is filled by a mixture of air and water and this region is referred to as the unsaturated zone or zone of acration.

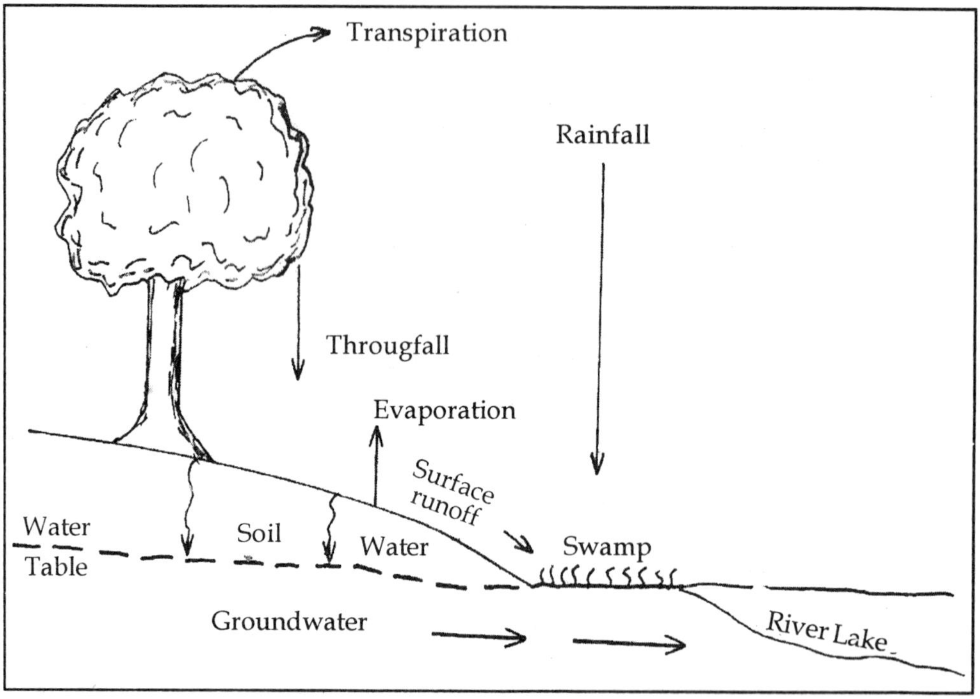

Fig. 4.1. Pathway of water near land surface (after Berner and Berner, 1987).

Groundwater flows underground until the water table intersects the land surface and the flowing water becomes surface water in the form of springs, rivers, swamps, and lakes.

Water coming into contact with rocks (and derived soils) reacts with primary minerals contained in them. The minerals dissolve to varying extents and some of the dissolved constituents react with one another to form new, or secondary, minerals. Dissolution is brought about mainly by acids provided by plants activity and bacterial metabolism, and the overall process is called chemical weathering. Besides biological factors, chemical weathering is also aided by physical processes that act to break up rocks and expose additional fresh mineral surface area to be attacked by weathering solutions which also aid chemical weathering. This is known as physical weathering and the dominant process is the fracturing, jointing of rocks by expansion due to heating accompanied by the freezing of water in cracks due to cooling. In tropical climate, due to heating during the day time (and/or during summer months) and low temperature in the night or wet weather during rainy season and freezing in high latitudes, physical weathering is important. In Himalaya such phenomena, is commonly noticed. Together chemical, biological, and physical weathering results in the breakdown of rock and the formation of soil.

Besides soil formation, chemical weathering also results in radical change in the composition of soil water and groundwater. These changes reflect both the composition of the primary minerals and the degree of biological activity acting to bring about mineral dissolution.

Chemical Weathering (minerals involved in weathering)

Chemical weathering of rocks (igneous, sedimentary and metamorphic) involves destruction of mineral constituents of which the rocks are composed. Rocks consists of primary, secondary and accessory minerals; the former is essential constituent of the rock, the secondary is derived from the primary by weathering and accessories are the mineral constituents formed during the birth of the rock (Table 4.1).

Table 4.1: Susceptibility of primary minerals to weathering in decreasing order

Mineral	Generalized chemical formula	Rock types	Main weathering reaction
Olivine	$(Mg, Fe)_2 SiO_4$	Igneous	Oxidation of Fe Cong. Diss. by acids
Pyroxenes	$Ca (Mg, Fe) Si_2O_6$ or $(Mg, Fe) SiO_3$	Igneous	Oxide of Fe cong. diss. by acids
Amphiboles	$Ca_2 (Mg, Fe)_5 Si_8O_{22} (OH)_2$ (also some Na & Al)	Igneous Metamorphic	Oxide of Fe cong. diss. by acids
Plagioclase feldspar	Solid solution between $Na Al Si_3O_8$ (albite) and $Ca Al_2 Si_2O_8$ (anorthite)	Igneous Metamorphic	Incong. diss. by acids
K-feldspar	$KAl Si_3O_8$	Igneous Metamorphic Sedimentary	Incong. diss by acids
Biotite	$K (Mg, Fe)_3 (Al SiO_3 O_{10}) (OH)_2$	Metamorphic Igneous	Incong. diss by acids oxide of Fe
Mucovite	$K Al_2 (Al Si_3O_{10}) (OH)_2$	Metamorphic Igneous	Incong. Diss. by acids
Quartz	SiO_2	Igneous, metamorphic, sedimentary	Resistance to dissolution
Calcite	$CaCO_3$	Sedimentary	Cong. Diss. by acids
Dolomite	$Ca Mg (CO_3)_2$	Sedimentary	Cong. Diss. by acids
Pyrite	$Fe S_2$	Sedimentary	Oxide of Fe and S
Gypsum	$CaSO_4.2 H_2O$	Sedimentary	Cong. Diss.by H_2O
Anhydrite	$CaSO_4$	Sedimentary	Cong. Diss.by H_2O
Halite	NaCl	Sedimentary	Cong. Diss. by H_2O

(Modified after Berner and Berner 1987)

In the present context weathering reactions are classified according to the nature of the attacking substance and whether the primary minerals simply dissolve or/and whether a portion of it reprecipitates to form a secondary mineral or minerals. Simple dissolution is referred to as congruent dissolution, and dissolution with reprecipitation of some of the components of the mineral is called incongruent dissolution. Commonly the substances that attack the constituent minerals are soil acids, dissolved oxygen, and water itself. Dissolved oxygen only attacks those minerals that contain reduced forms of elements, principally iron and sulphur, and that undergo oxidation to form new minerals.

Although most minerals are attacked mainly by soil acids, a few very soluble ones simply dissolve in water and such minerals may reprecipitate under arid conditions.

Silicate Weathering

Of all mineral groups, silicates have received the most attention in weathering studies because they make up the most abundant rock types. During weathering, soil solutions penetrate through permeable clay layers right to the bare surfaces of the primary silicate mineral grains and there reacts with them. Dissolution does not occur at all places on the surface so as to produce general rounding of the grains, but instead affects only those portions of the surface where there is excess energy such as at outcrops of dislocations or lattice defect in crystals.

In the case of silicate minerals, congruent dissolution is rare and confined only to relatively iron-free olivine, amphiboles, and pyroxenes. Most other silicate minerals, especially those containing aluminum, dissolve incongruently with the consequent formation of iron oxides and /or clay minerals, fine-grained aluminosilicates and includes kaolinite, smectite, and vermiculite as well as other minerals, such as chlorite.

The silicate weathering consists solely of attack by carbonic (and sulphuric) acids (Garrels, 1967). The organic acids really do much of the attacking but they disappear. It is assumed that all aluminum liberated by feldspar dissolution was precipitated to form a secondary mineral (in this case Kaolinite). However, all Al reprecipitated may not always form kaolinite but other secondary minerals, gibbsite $Al(OH)_3$, and smectite may form. Except for localized redistribution accompanying chelate transport, aluminum does not migrate any appreciable distance in solution and normally accumulates in soils as weathering proceeds. In fact the change in the ratio of other elements to aluminum in soils is often used as a measure of the degree of removal by weathering (e.g. Goldich, 1938). Iron, because of its insolubility in the presence of dissolved O_2, also accumulates in soils, as ferric oxides.

The most abundant silicate minerals in the earth's crust, which also readily undergoes weathering, is plagioclase feldspar. The Na - plagioclase (albite) had reacted with carbonic acid:

$$2H_2 CO_3 + 9H_2O + 2Na\,Al\,Si_3O_8 \rightarrow Al_2Si_2O_5\,(OH)_4 + 2Na^+ + 2HCO^-_3 + 4H_4SiO_4$$

The H_2CO_3 forms by the reaction

$$H_2O + CO_2 \rightarrow H_2\,CO_3$$

Silicate Weathering: Soil Formation

Soil can be defined as "a complex system of air, water, decomposing organic matter, living plants and animals, in addition to the residues of rock weathering, organized into definite structural patterns as dictated by the environmental conditions (Loughnan, 1969). Topography, climatic conditions, groundwater regimes and organic activity (primarily vegetal) govern the genesis of soils. On the hillslopes, the soil cover is thinner as a result of denudation, while on flat ground it is thicker. Formation of soil (Pedogenesis) is a long term process stretching over periods ranging from a few hundred to a few thousand years. The role of climate, particularly higher temperature and rainfall, is very critical in the formation of soils. This is evident from the thickness and nature of soil profile in relation to latitudes, that is climatic zones. Water is the controlling factor in determining the chemical composition of soils, as it stimulate the cation exchange on soil colloids and regulating the rate of production of plants.

A number of processes are involved in the genesis of soil. Under cold humid climatic conditions, organic acids are produced during humification. These acids decompose the minerals, the hydrogen ions of the acid replacing the cations K, Na, Ca, Mg. The resulting compounds are leached down by interstitial water from the upper layer leaving it deficient in bases colloids and oxides of Fe and Al, but richer in SiO_2. The base-deficient acidic soil is called podsol and the process of formation is known as podsolization (Fig.4.2). Organic matter is important as it is a source of carbonic and organic acids and chelating compounds that react with silicate minerals.

The principle acids produced by soil microorganisms are carbonic and sulphuric acids and a whole host of various organic acids. Carbonic acid (H_2CO_3) is produced by the oxidation of organic matter (CH_2O) by microbes to CO_2

$$CH_2O + O_2 \rightarrow CO_2 + H_2O$$

This CO_2 then combines with water to form carbonic acid, which further partly dissociates to H^+ and HCO_3-

$$CO_2 + H_2O \rightarrow H_2\,CO_3$$

$$H_2 CO_3 \rightarrow H^+ + HCO_3$$

Sulphuric acid ($H_2 SO_4$) is produced by the bacterially catalyzed oxidation of sulphide minerals and, in soils developed on rocks with high sulphide mineral contents.

In the hot, wet climate belts characterized by abundance of water, the bacteria are very vigorous and destroy humus considerably so that the soil becomes deficient in organic acids. In the deficiency of organic acids, sesquioxides of Fe and Al are not converted, dissolved and removed by leaching. Consequently, the sesquioxides Al_2O_3 and Fe_2O_3 (with FeO) accumulate in the soil as red and yellow clays and pisolitic nodules. The result in the formation of laterosol at low fertility. The process is known as laterization (Fig.4.2).

Carbonate Weathering

Compared to silicates, carbonate minerals weather far more easily. Dissolution is normally congruent and the overall reaction, accomplished by carbonic acid (H_2CO_3) attack forms calcite:

$$H_2CO_3 + CaCO_3 \rightarrow Ca^{++} + 2HCO^{-}_3 \qquad \text{..........(i)}$$

and for dolomite

$$2H_2CO_3 + CaMg\,(CO3)_2 \rightarrow Ca^{++} + Mg^{++} + 4\,HCO_3^- \qquad \text{.............(ii)}$$

Carbonate minerals are not as abundant as silicate minerals but they exert a dominant influence on groundwater and surface water composition. Most of the dissolved Ca^{++} and HCO_3^- in river water on a worldwide basis, arises from carbonate dissolution via reactions (i) and (ii).

Calcium carbonate dissolution, under certain circumstances, can be followed by reprecipitation. In soil water, and groundwater where carbonic acid concentrations are high, the resulting concentrations of Ca^{++} and HCO_3^- can build up to high values. If the water then encounters conditions where degassing of dissolved CO_2 occurs, the water may become supersaturated with respect to $CaCO_3$. Carbonic acid constantly maintains equilibrium with dissolved CO_2:

$$H_2CO_3 \leftrightharpoons H_2O + CO_2 \qquad \text{....................(iii)}$$

and if CO_2 is lost from solution, reaction (iii) shifts to the right, using up H_2CO_3 and consequently, reaction (i) shifts to the left to replace H_2CO_3. If Ca^{++} and HCO_3- are present in high enough concentrations, $CaCO_3$ precipitation may then take place.

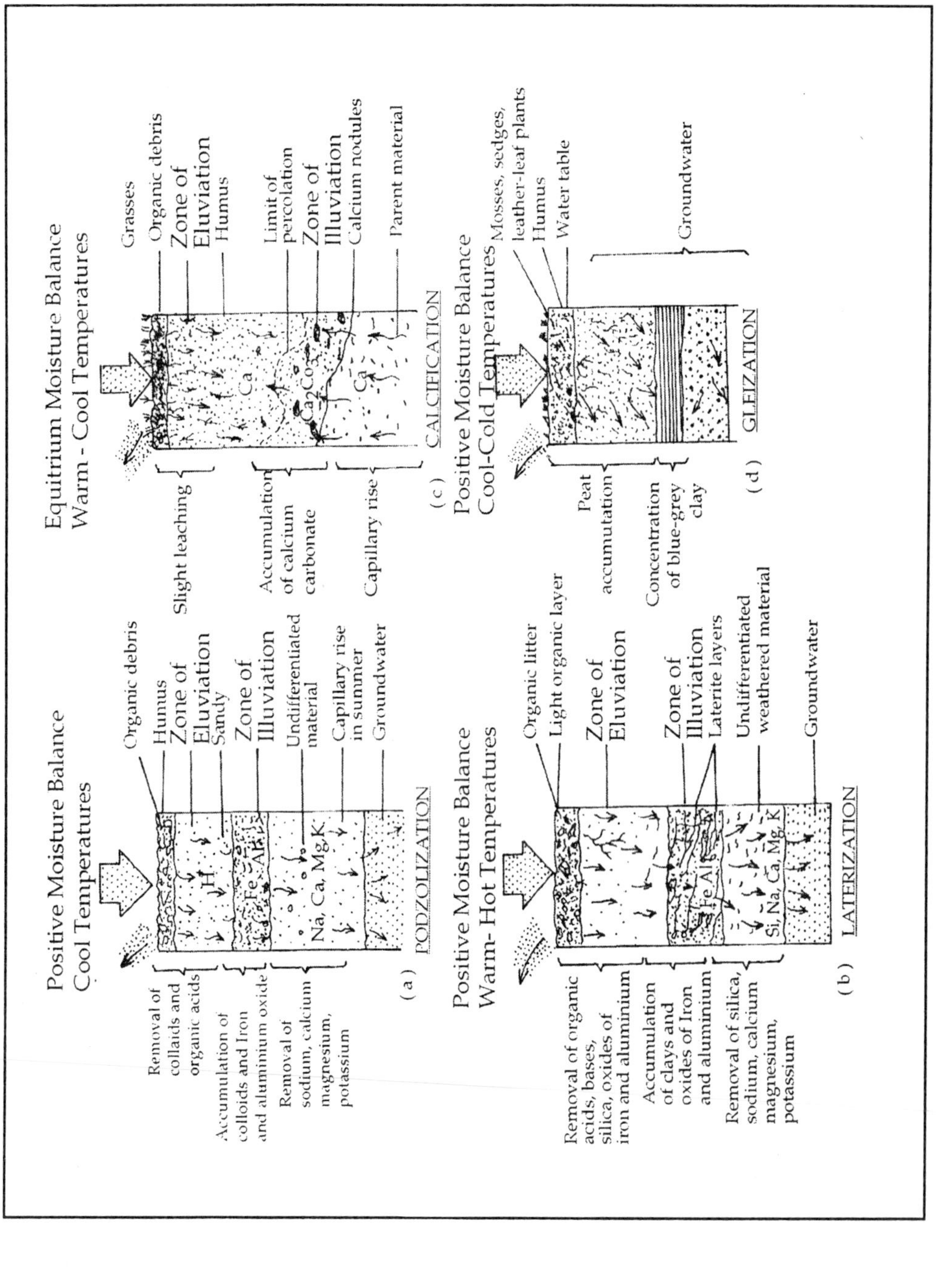

Fig. 4.2. Processes involved in pedogenesis of soil (based on Marsh and Dozier, 1981).

Sulphide Weathering

Sulphide minerals occur in minor quantities in different rock types and in greater concentrations in some ore deposits. The most abundant mineral, by far, is pyrite (FeS_2), which is found mainly in organic-rich, fine-grained sedimentary rocks black shales, and in coal. Upon exposure to dissolved oxygen during weathering, pyrite (and other sulphides) are chemically unstable and rapidly undergo oxidative decomposition. This decomposition results in the production of sulphuric acid which is used for further weathering of silicate and carbonate minerals. Sulphuric acid forms as follows:

$$4\,FeS_2 + 15\,O_2 + 8\,H_2O \rightarrow 2\,Fe_2O_3 + 8\,H_2SO_4 \quad \text{(iv)}$$

$$H_2SO_4 \rightarrow 2H^+ + SO_4^- \quad \text{(v)}$$

Almost always the oxidation is catalyzed by bacteria (Stumm and Morgan, 1981).

Calcareous rocks that also contain pyrite do not weather to produce highly acid water like that of coal-mining localities. Instead the sulphuric acid is neutralized by $CaCO_3$.

$$4FeS_2 + 15O_2 + 8\,H_2O \rightarrow 2Fe_2O_3 + 16\,H^+ + 8\,SO_4^-$$

$$16\,H^+ + 16\,CaCO_3 \rightarrow 16\,Ca^{++} + 16\,HCO_3^-$$

which, added together, give

$$4FeS_2 + 15\,O_2 + 8\,H_2O + 16\,CaCO_3 \rightarrow 2Fe_2O_3 + 16Ca^{++} + 8\,SO_4^- + 16\,HCO_3^-$$

In this way Ca^{++} - SO_4^- - HCO_3^- groundwaters and stream waters can arise. In arid and semiarid regions where evaporative concentration of soil waters is common, the Ca^{++} and SO_4^-concentrations often reach the point where the precipitation of gypsum, $CaSO_4 \cdot 2H_2O$ takes place.

Sulphuric acid is also partially neutralized by silicate minerals. It is this neutralization, in fact, that constitutes one of the more important weathering reactions for silicates. For example, for albite:

$$H_2SO_4 + 9H_2O + 2\,Na\,Al\,Si_3O_8 \rightarrow Al_2Si_2O_5\,(OH)_4 + 2Na^+ + SO_4^- + 4\,H_4\,SiO_4$$

□□□

Chapter 5

WEATHERING AND WATER CHEMISTRY EXAMPLES

LAKES OF KASHMIR

Dal Lake

The Dal lake is situated at an altitude of 1583 m above mean sea level (MSL) in the Kashmir valley. The lake is located between 34° 18' N latitude and 74° 91' E longitude having total area of 16.4sq kms (NRSA, 1996) and catchment area 316 km^2 which is about 16 times more than the lake area. The maximum depth of the lake is about 3.5 meters (ENEX, 1978) and average depth of lake is 1.5 to 2.0 meters. A review of available data indicates that the lake has undergone significant changes in its dimensions since the 15th century. The creation of Moghul Gardens along its periphery, construction of roads, bunds and island have fragmented the vast expanse of water into small basins (Gagribal, Bod-Dal, Hazratbal and Nagin). Currently, of the total lake area (21 km^2) only about half (11.7 km^2) is under open water and 3.9 km^2 under floating gardens, 2.5 km^2 marshy land and 1.41 km^2 land area (Kanth and Rather, 2001). According to recent study the floating gardens cover an area of 7.5 km^2 within the Dal Lake. The eastern side of the lake is flanked by Mahadev Range and Southern by Kohi Suliman Range (Fig.5.1).

The lake is fed by a number of nallahas draining the catchment. However, the main source of water is Talbal nallah, coming from Mansar Lake situated in the higher reaches of Dachigam National Park. The most conspicuous drainage feature around Dal Lake is occurrence of loosing streams. These signify their ending in highly porous sediments. According to the geological

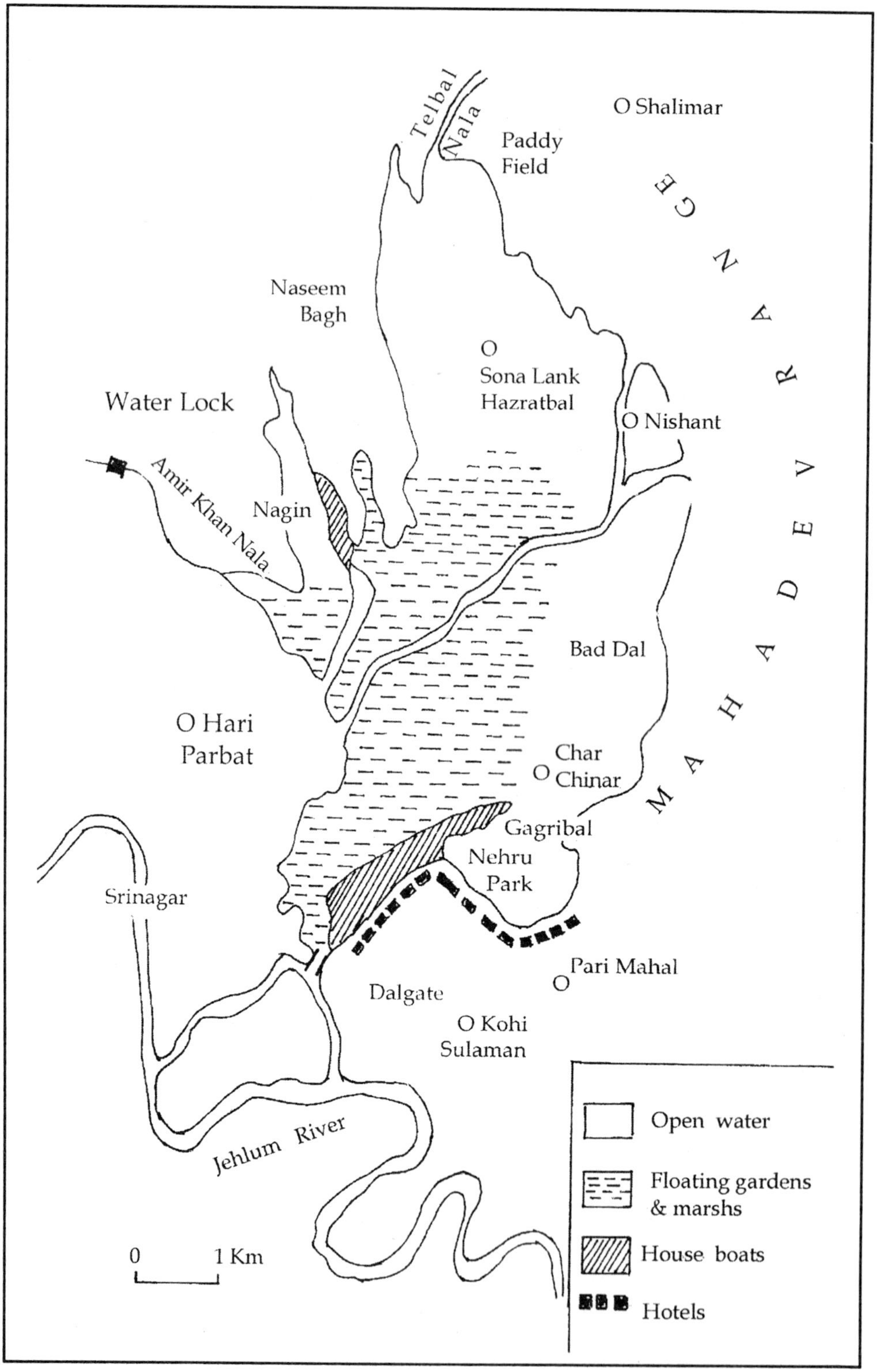

Fig. 5.1. Outline map of Dal Lake showing configuration of lake, Marshy Land, Hotels etc.

knowledge available, these loosing streams do end up in the Karewas. These streams are of ephemeral (seasonal) nature. The runoff ultimately infiltrates in the Karewas due to their loosing nature. This goes as sub-surface water and recharges the Dal Lake. The lake, receives a considerable amount of water as a result of groundwater recharging from the eastern and northern sides. The lake is flushed through two outlets one being the Dalgate, through which the lake discharges water into a tributary of river Jhelum. The other outlet being a Nagin Lake where a small canal, Nallah Amir Khan connects the Dal Lake with Anchar Lake. The catchment area of the Dal Lake consists of Karewas (sandy, silty and pebbly conglomerates), Jurassic and Triassic limestone, Permo-carboniferous sandstone, clay-shale and limestone, Panjal Trap and alluvium. According to Tripathi and Safaya (1998) 40% of Dal catchment area is occupied by Karewas and alluvium comprising sand, silt and clay which are susceptible to easy erosion and transportation causing greater sediment inflow into the lake.

The water quality parameters for the year 1977 and 1996 (Tables 5.1 and 5.2) show variation for the four drainage basins of the Dal Lake. A significant variation in pH and total phosphorus and Ammonical Nitrogen in the year 1996 indicate increasing pollution in the lake. The high nutrients indicate anthropogenic source and increase in pH may suggest respiration in the basin due to biomass degradation and release of CO_2. However, the investigation which is being carried out in the recent years indicates pH ranging from 7.4 to 8.5, electrical conductivity (E.C) from 110-200µs, DO 7-16mg/l, total alkalinity 70-120 mg/l, nitrate-N (165-395µg/l), ammonical nitrogen (10-65 µg/l), phosphorus –P(45-116µg/l). The lower nutrients level in water may indicate algal bloom period, which may have been utilized by the organisms for growth and aggregation. This is a well known fact, that the nitrogen and phosphorus are the most important constituents of the organic matter and are hence required by the biota. The other parameters are on increase indicating increasing pollution in the lake. The increase in the algae, phytoplankton and other biomasses together with increase in siltation rate are responsible for the shrinkage of Dal Lake. The dense growth of microphytes, causes considerable water loss through evatranspiration, and adds huge amounts of organic matter which in turn raises the BOD (Zutshi and Gopal, 2000). In addition deforestation, overgrazing and agricultural activity have resulted in huge soil erosion and subsequent deposition into the lake basins. It is estimated that 36 million m^3 per year of silt is discharged into the lake through the Telbal Stream which drains the upper reaches of Dachigam (ENEX, 1978; Zutshi and Gopal, 2000). The sedimentation rate of 117000 sq meter/yr has been estimated (Kanth and Rather, 2001).

Table 5.1: Water quality parameters of Dal Lake, Kashmir(1977)

Parameters	Hazratbal	Bod Dal	Gagribal	Nagin
pH	7.7-9.5	7.4-9.5	7.5-9.5	7.7-9.5
E.C (μs/cm at 25°C)	290-485	279-431	302-543	305-490
D.O	2.2-12.0	5.5-11.5	4.5-10.5	0.8-10.0
Tot. alkalinity	70-120	2.2-120	70-125	80-134
Ca	19-39	18-39	21-125	19-39
Mg	8-13	8-13	8-12	8-12
Na	4.2-5.6	3.6-5.0	3.3-3.5	4.6-7.2
K	0.3-1.6	0.4-1.6	0.3-1.2	.4-1.3
SiO_2	1.5-3.6	0.7-3.3	0.2-2.9	0.2-2.2
Nitr. Nitrogen	80-650	95-691	80-603	90-632
Ammo. Nitrogen	20-30	5-28	9-26	5-25
Total Phos.	62-623	65-620	70-500	90-873

Taken from Trisal (1987)
All parameters in mg/l

Table 5.2: Water quality parameters of Dal Lake, Kashmir(1996)

Parameters	Hazratbal	Bod Dal	Gagribal	Nagin
pH	8.1-9.8	7.5-9.3	8.2-9.6	8.0-9.7
E.C (μs/cm at 25°C)	155-409	130-483	116-336	204-423
D.O	3.4-10.7	2.7-10.4	2.4-10.2	1.2-10.4
Tot. alkalinity	76-173	56-208	17-136	81-269
Ca	14.1-17	14.6-61	17-26.2	5.2-52
Mg	0.4-15.9	0.6-9.2	0.7-8.7	1-18.3
Na	5-38	4-40	4-34	10-62
K	1-22	1-12	0-10	2-30
SiO_2	02-19.5	0.3-14.9	0.3-14	0.3-32.5
Nitr. Nitrogen	7-255	4-503	2-243	5-599
Ammo. Nitrogen	37-726	26-627	24-705	43-1357
Total Phos.	151-423	136-8106	136-5060	121-4182

Taken from Kundangar and Sarwar (1997)
All parameters in mg/l

The flow of untreated sewage from human settlements, hotels, houseboats, dunga boats is a regular feature in the Dal Lake. The estimate shows that 16.5 tons of P and 364.5 tons of N are drained annually into the lake with a net balance 5.5 t P and 88.9 t N (ENEX, 1978) resulting in the deterioration of water quality of lake since many years.

Mansar Lake

The Mansar lake of Jammu is the deepest fresh water lake at an altitude of 660 m from msl and located at an anticline axis of a large fold (Surin-Mastgarh anticline) (Singh and Sharma, 1999) (Fig.5.2). It is bound by 32^0 48' N latitude and 75^0 23' E longitude having an area of 0.59 km^2 and circumference of 3020 m. The maximum depth of the lake is 38.25m in the centre with maximum length 1204 m and width 645 m. The average annual rainfall is 1500 mm, and lake water level varies between 1m to 1.5 m in a year. There is no perennial stream flowing into the lake from the catchment. The lake is thus fed essentially by rainwater entering as overland flow and through two ephemeral streams (Fig. 5.3 a,b). Poor population and lack of inlet nallahs saved the lake from getting polluted.

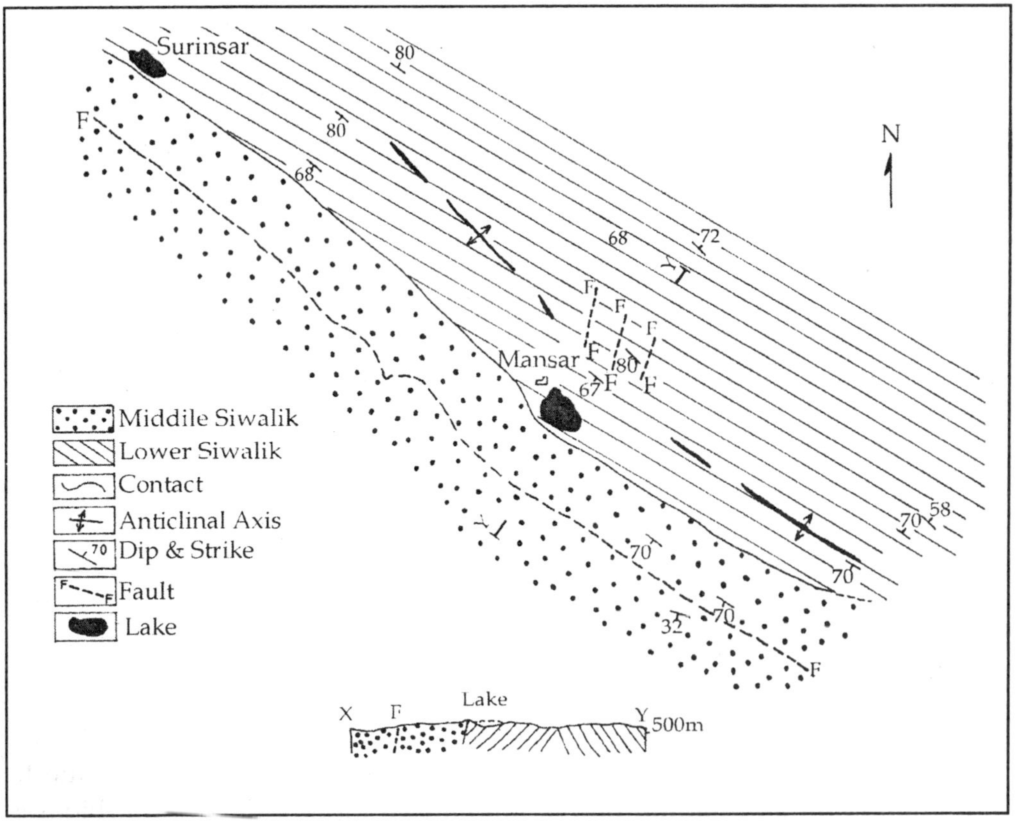

Fig. 5.2. Geological and tectonic map of Mansar and Surinsar Lake showing lithology, anticlinal axis and faults (after Singh and Sharma, 1999).

The Mansar Lake catchment is comprised of fine grained sandstone interlayered siltstones, mudstones and clays of lower Siwalik Formation (Fig.5.2). The calcareous cement is commonly observed in sandstone microscopically. The folding faulting of lower Siwalik has resulted in crushing of rocks and these crushed rocks form the porous and permeable zone for recharge of the lakes (Rai *et al.,* 2006).

The lake is alkaline in nature with pH ranging from 7.62-7.91 with an average of 7.75 during the mixing (Winter Overturn) of the lake, and during stratification pH changes from 8.44 to 8.77 with an average value of 8.65 (Table 5.3 and 5.4). Investigations have shown that the temperature decreases with depth from April to November in Mansar Lake and during this period, the upper relatively warm layer ranging up to 6 m depth is epilimnion and below this, temperature decreases rapidly between 6 to 10 m depth and this zone of maximum temperature change is thermocline. Below this zone at 10 m depth, there is relatively cold layer, which is hypolimnion. The stable stratification that persists from April to November tends to become unstable during early winter (i.e. December), due to lowering of air temperature resulting in decrease of water temperature of epilimnion and thermocline zone due to net heat loss from the lake. The development of uniform temperature conditions initiates the process of convection, which coupled with the increased action of wind, results in the deepening of surface layer (sinking of thermocline) and complete mixing of lake takes place. This period of vertical mixing of Mansar Lake is "winter overturn" when lake is almost completely mixed. The Mansar Lake, therefore, undergoes two stage process namely complete mixing and a stratified stage of minimal vertical mixing (Rai *et al.,* 2006).The higher pH value of surface water during stratification due to enhanced photosynthesis and organic matter production and consequent relative depletion in CO_2. The pH decreases with depth (epilimnion to hypolimnion) during period of lake stratification and it is uniform from surface to bottom during winter overturn.

A low pH in hypolimnion is due to decomposition of organic matter under oxygen deficient conditions enhancing CO_2 in water and during mixing in winter uniform pH is recorded. Increasing trend of conductivity with depth is due to mineralization of organic matter under reducing conditions prevailing in the hypolimnion zone hence depth water shows greater conductance in comparison to surface.

The dissolved oxygen (D.O) 7.5-8.0 mg/l on average in surface water depletes in winter very much resulting in fish mortality often reported from Mansar Lake. The lower value of D.O in winter in surface water is due to mixing of lake water and during summer season photosynthesis of biota is responsible for higher D.O (Ahmed et al, 2003).

Fig. 5.3 (a) Mansar Lake showing lower Siwalik Hills in its catchment and settlement along the bank.

Fig. 5.3 (b). Panormic view of Mansar Lake showing vegetation on Siwalik hills.

Table 5.3: Chemical Characteristics of Mansar Lake water during mixing.

Sample No.	pH	E.C. µs/cm	D.O.	CO_3^{2-}	HCO_3^{-2}	Cl^-	SO_4^{-2}	PO_4^{-3}	SiO_2	Na^+	K^+	Ca^{+2}	Mg^{2+}	T.H.	TDS	
			....mg/l....													
Man-1	7.80	255	8.9	0.0	122.0	12.78	1.0	0.018	2.42	10.79	2.07	63.0	8.0	190.3	150	
Man-2	7.62	255	9.1	0.0	122.0	12.78	0.5	0.062	2.13	12.81	2.07	60.0	7.0	178.7	150	
Man-3	7.72	256	8.8	0.0	122.0	11.36	1.0	0.029	2.00	12.81	2.07	63.0	8.0	190.3	150	
Man-4	7.72	255	8.3	0.0	122.0	11.36	1.0	0.000	2.09	12.48	2.07	57.0	6.0	167.3	150	
Man-5	7.70	260	8.1	0.0	146.4	11.36	0.5	0.015	2.21	12.81	2.07	57.0	8.0	175.3	150	
Man-6	7.68	256	8.3	0.0	146.4	11.36	1.0	0.000	2.09	13.49	2.07	66.0	7.0	193.7	150	
Man-7	7.80	257	8.4	0.0	122.0	12.78	0.5	0.018	2.13	13.49	2.07	63.0	8.0	190.3	150	
Man-8	7.74	257	8.2	0.0	146.4	12.78	0.5	0.045	2.16	13.83	2.42	60.0	8.0	182.8	150	
Man-9	7.74	258	8.7	0.0	146.4	12.78	0.5	0.018	2.18	13.49	2.42	60.0	7.0	178.7	150	
Man-10	7.74	259	8.2	0.0	122.0	12.78	0.5	0.043	2.15	13.49	2.42	60.0	7.0	178.7	150	
Man-11	7.70	260	7.4	0.0	146.4	12.78	0.5	0.013	2.09	13.49	2.42	66.0	6.0	189.5	150	
Man-12	7.68	255	10.4	0.0	122.0	11.36	0.5	0.456	1.85	13.49	2.07	57.0	9.0	179.4	150	
Man-13	7.66	255	8.3	0.0	122.0	11.36	0.5	0.000	1.94	12.81	2.07	57.0	6.0	167.0	150	
Man-14	7.70	257	11.2	0.0	122.0	11.36	0.5	0.035	1.79	13.15	2.07	57.0	9.0	179.4	150	
Man-15	7.75	260	9.0	0.0	146.4	12.78	0.5	0.018	1.79	12.81	2.07	60.0	7.0	178.7	150	
Man-16	7.84	253	9.9	0.0	146.4	12.78	0.5	0.026	1.46	12.81	2.76	57.0	6.0	167.0	150	
Man-17	7.84	255	8.1	0.0	122.0	11.36	0.5	0.015	1.68	13.49	2.76	63.0	6.0	182.0	150	
Man-18	7.91	274	8.3	0.0	122.0	11.36	0.5	0.026	1.80	13.49	2.76	57.0	6.0	167.0	140	
Man-19	7.80	237	8.3	0.0	146.4	11.36	0.5	0.000	1.55	12.81	2.76	57.0	6.0	167.0	140	
Man-20	7.83	238	8.6	0.0	146.4	12.78	0.5	0.029	1.52	12.14	2.07	54.0	7.0	163.7	140	
Man-21	7.82	236	8.8	0.0	122.0	12.78	0.5	0.018	1.82	12.14	2.42	57.0	6.0	167.0	140	
Man-22	7.84	237	7.9	0.0	122.0	12.78	0.5	0.013	1.77	12.14	2.42	57.0	9.0	179.4	140	

Table 5.4: Chemical Characteristics of Mansar Lake during stratification

Sampl e No.	pH	EC μs/cm	ORP (mV)	DO	TDS	CO_3^{2-}	HCO_3- mg/l	SO_4^{2-}	$Cl^?$	PO_4^{3-} (μg/l)	SiO_2	Ca^{+2}	Mg^+_2	Na^+ (mg/l)	K^+	Total Hardness
Man-1	8.44	190	286	8.0	120	6.0	103.7	10	12.78	1.5	1.09	27.0	6.4	12.14	2.76	93.79
Man-2	8.77	190	286	8.5	120	6.0	103.7	10	12.78	1.5	1.03	27.6	3.5	12.14	3.11	83.35
Man-3	8.76	190	288	8.9	120	6.0	91.5	10	11.36	3.2	0.97	26.7	3.4	12.81	3.11	80.69
Man-4	8.66	190	287	9.5	120	9.0	97.6	10	12.78	3.2	0.71	29.1	3.7	13.49	3.46	87.92
Man-5	8.65	190	287	9.6	120	9.0	79.3	15	11.36	6.2	0.77	25.8	3.5	12.81	3.46	78.85
Man-6	8.74	180	288	9.8	120	6.0	85.4	15	12.78	6.2	0.84	26.1	3.6	12.81	3.46	80.01
Man-7	8.69	190	296	9.4	120	6.0	91.5	10	12.78	3.2	1.09	25.8	3.6	12.81	3.46	79.27
Man-8	8.71	190	300	9.0	120	9.0	79.3	15	12.78	7.8	1.03	25.8	4.1	12.14	3.80	81.32
Man-9	8.60	190	302	9.0	120	6.0	85.4	10	12.78	0.0	1.03	23.1	4.4	12.14	3.46	75.81
Man-10	8.60	190	302	9.1	120	6.0	85.4	10	14.20	3.2	0.84	26.1	3.6	10.78	3.46	80.01
Man-11	8.66	190	302	8.7	120	6.0	85.4	15	12.78	1.5	0.77	27.0	3.5	10.78	3.46	81.85
Man-12	8.60	190	308	8.4	120	9.0	91.5	5	12.78	3.2	0.90	26.7	3.7	11.46	3.46	81.92

The TDS is high 147 mg/l during mixing of lake and it is low 120 mg/l during stratification accordingly alkalinity is also high 2.16 meq/l in case of former and low 1.59 meq/l in case of latter (Tables 5.3 and 5.4). The variation in TDS is due to Ca^{2+} precipitation during stratification of lake. Calcium and magnesium are major cations in the Mansar Lake, together accounting for about 89.91equi % and CO_3^{2-} + HCO_3 85.84% during winter when mixing of lake takes place and $Ca^{2+} + Mg^{2+}$= 77.57% and $CO_3^{2-} + HCO_3^- = 74.23\%$ during summer stratification. However, the major cations in Mansar Lake for both the seasons are, in decreasing order, $Ca^{2+} >> Mg^{2+} > Na^+ > K^+$ anions in the same order are $HCO_3^- >> SO_4^{2-} > Cl$. Therefore, Ca^{2+} is the dominant cation and bicarbonate (HCO_3^-) is the dominant anion in the lake water and general chemical composition of lake water is close to $[HCO^{3-}]$ ³ 3 $[Ca^{2+}]$. This typically indicates high dissolution of calcite even though lower Siwaliks are poor in carbonate rocks, and silicates (clay-shales-sandstone) predominates.

The lake water chemistry is primarily controlled by the weathering and erosion of lithology. The plot of TDS versus weight ratio of Na/(Na + Ca) on Gibbs (1970) diagram, show predominance of weathering as a major source of solute to the water of the lake (Fig.5.4). Since the weathering of different parent rocks (e.g. of carbonate and silicate) yields different combinations of cations and anions in solution (Garrels and Mackenzie, 1971; Stumm, 1992), a probe into the mechanism of weathering (dissociation) of these rocks would explain the ionic concentration in lake water. The high concentration of elemental ions in Mansar Lake is due to the congruent weathering of carbonate and incongruent weathering of silicate rocks. Weathering of carbonates is the major source of Ca^{2+} and Mg^{2+} dissolution is fresh water and even though they (rocks) occur in much subordinate amounts at the source area, dominates the water composition (Berner and Berner, 1987).

In Mansar Lake same situation can be observed. The plot of water chemistry on Piper diagram (Fig.5.5) shows that calcium is the dominant cation and bicarbonate is the dominant anion in lake water and that it belongs to Ca- Mg -HCO_3 - CO_3 facies. However, HCO_3^- is produced by weathering of both silicate and carbonate minerals. Zhang and others (1995) observed that the stoichiometric relationships of Ca^{2+}/HCO_3^- and $Ca^{2+} + Mg^{2+}/HCO_3^-$ in case of carbonate weathering is 0.50 (in molar ratio). In Mansar Lake, it is 0.678 and 0.81, respectively in winter and 0.45 and 0.55 in summer samples. These values are close to the one reported by Zhang and others. The Ca^{2+}/HCO^{3-} and $Ca^{2+} + Mg^{2+}/ HCO^{3-}$ equivalent ratio in lake water is 1.2 and 1.37and 0.90 and 1.12, for winter and summer collections, respectively.

The elevated ratios indicates silicate weathering in addition to carbonate releasing ions to the water. The stoichiometry of carbonate weathering reaction demands that carbonate derived Ca and Mg should be equal to the carbonate derived bicarbonate (Singh and Hasnain, 1998). In a plot of (Ca+ Mg) vs

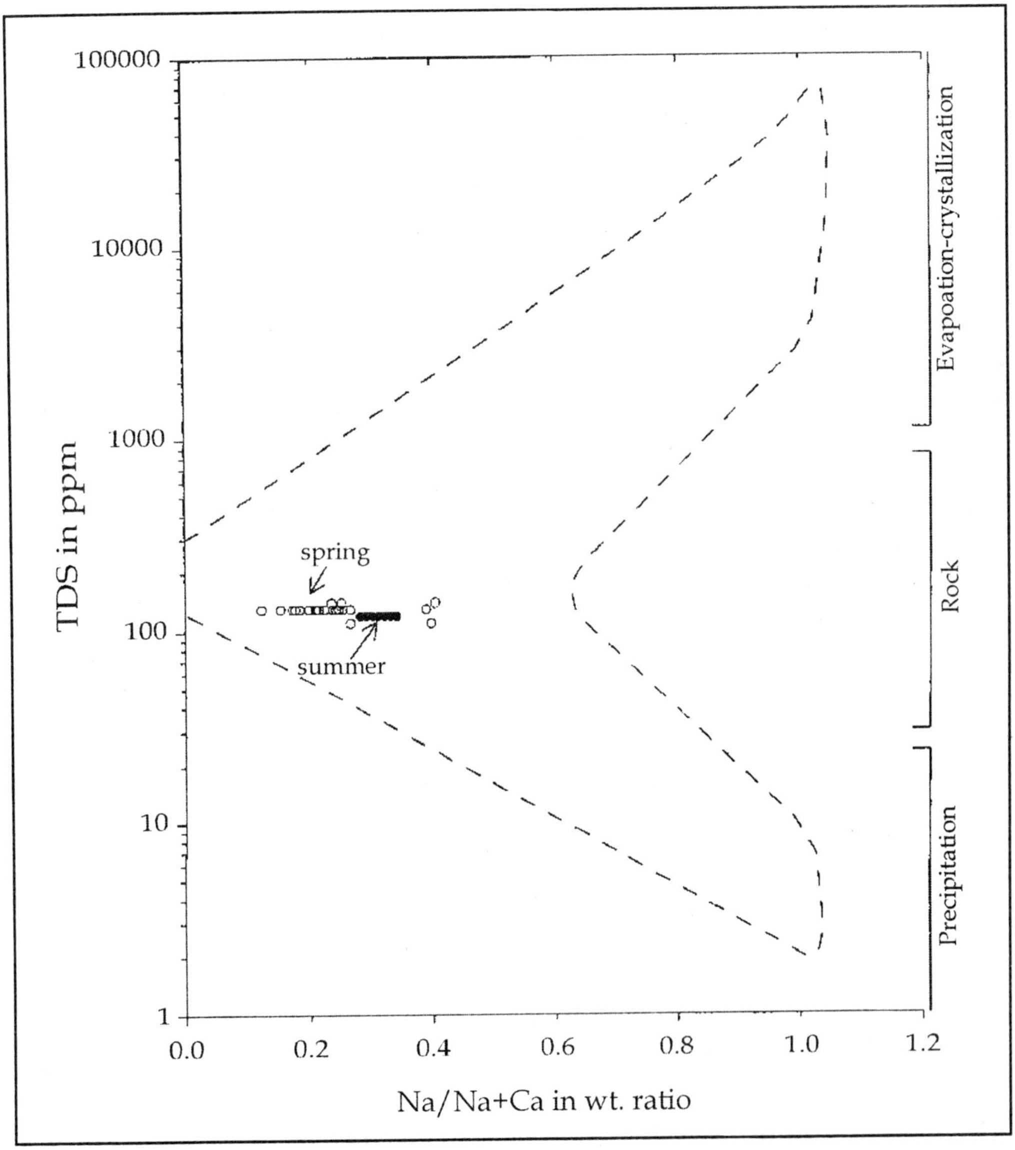

Fig. 5.4. Gibbs (1970) diagram depicting predominance of rock weathering in the Mansar catchment area.

HCO_3 (Fig.5.6a) most of the samples fall close to the equiline 1:1, where summer samples of surface water, when the lake is stratified, falls on the line and winter samples when lake is mixed, fall above the equi-line and few samples which are in excess of Ca+Mg in water are balanced by SO_4 and Cl. In a plot of (Ca+Mg) vs Tz^+ (total cations) the samples of both the seasons fall below the equiline in a linear trend (Fig.5.6b). Therefore the major contribution of (Ca + Mg) to the total cations (Tz_+), the relatively high (Ca + Mg) / (Na + K) ratio 4.26 in winter and 2.70 in summer and low (Na + K) / Tz^+ ratio 0.27 and

0.19, respectively indicate that the calcareous sandstone and mudstone of the Siwalik Group are the main source of lake water. Further, the (Ca + Mg)/(Na + K) molar ratio in the silicates of the upper crust is 1.0 (Taylor and McLennan, 1985), and in Mansar Lake it is 2.0. The higher ratio in the lake indicates a non-silicate source. This study is consistent with the conclusions arrived at on the Himalayan rivers (Sarin and others, 1989; Abbas and Subramanian, 1984; Pande and others, 1994; and Gale and France-Lanord, 1999). This is further evident from the molar ratio Ca/Na= 1.78, Mg/Na = 0.34 and HCO_3/Na = 3.20 which are much lower than those of the waters that drain carbonate lithology and are little higher than those that drains silicate rocks (Stallard, 1980; Meybeck, 1987, Negrel *el al.*, 1993). Therefore, these ratios may support both carbonate and silicate source contributing major elements to the water of the Mansar Lake. The Ca/Na molar ratio of the average crustal sedimentary rocks is close to 0.6 (Taylor and McLennan, 1985) and due to higher solubility of Na over Ca, lower Ca/Na molar ratio are expected in the dissolved loads of rivers draining silicate rocks. Therefore, the higher ratio of Ca/Na in Mansar Lake indicates non-silicate source, which contribute Ca in the basin.

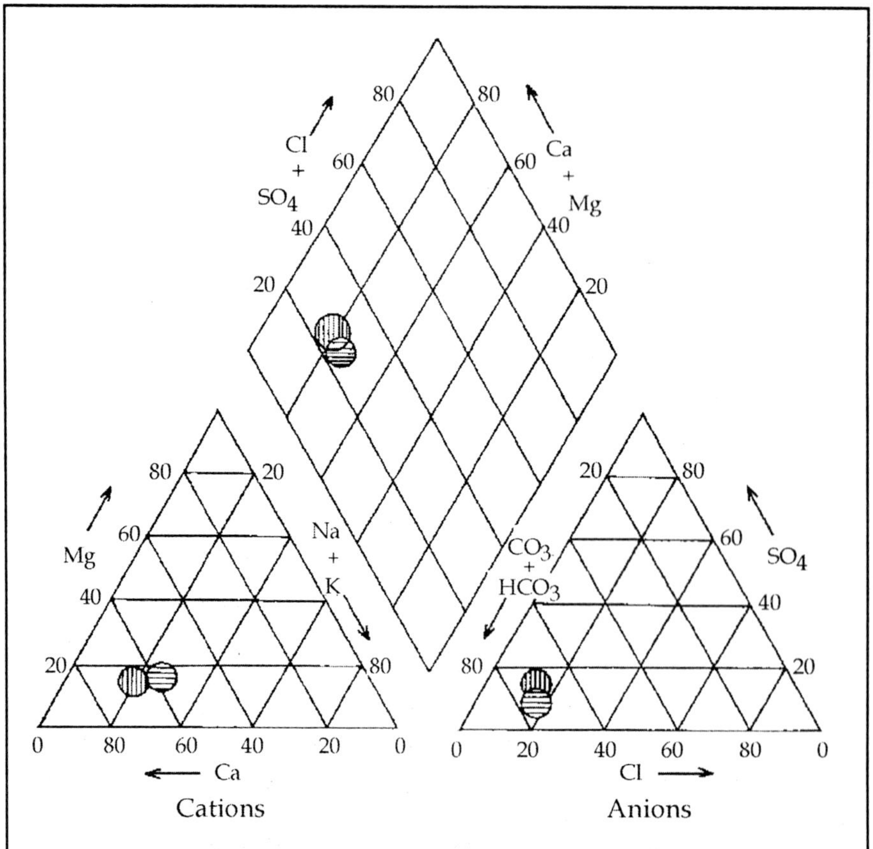

Fig. 5.5. Piper diagram showing Ca^{2+} as the dominant cation and HCO_3^- as anion making water Ca^{2+} - $HCO3^-$ type.

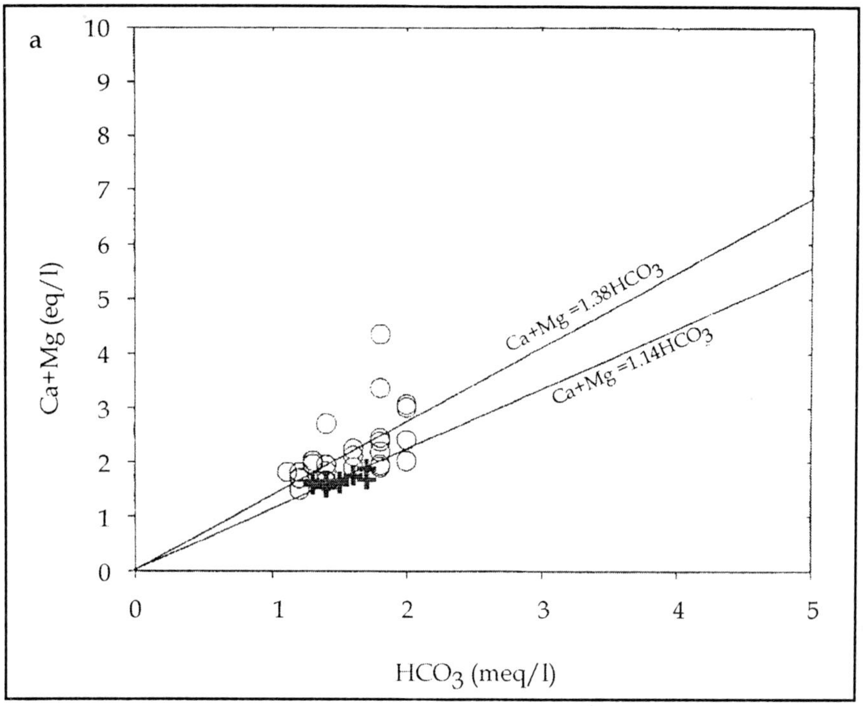

Fig. 5.6 (a) Scatter diagram : (a) plot of $Ca^{2+} + Mg^{2+}$ vs $HCO3^-$ of Mansar Lake showing summer data (+) plot close to the equiline 1:1 whereas winter data (O) when the lake is mixed, fall above the equiline.

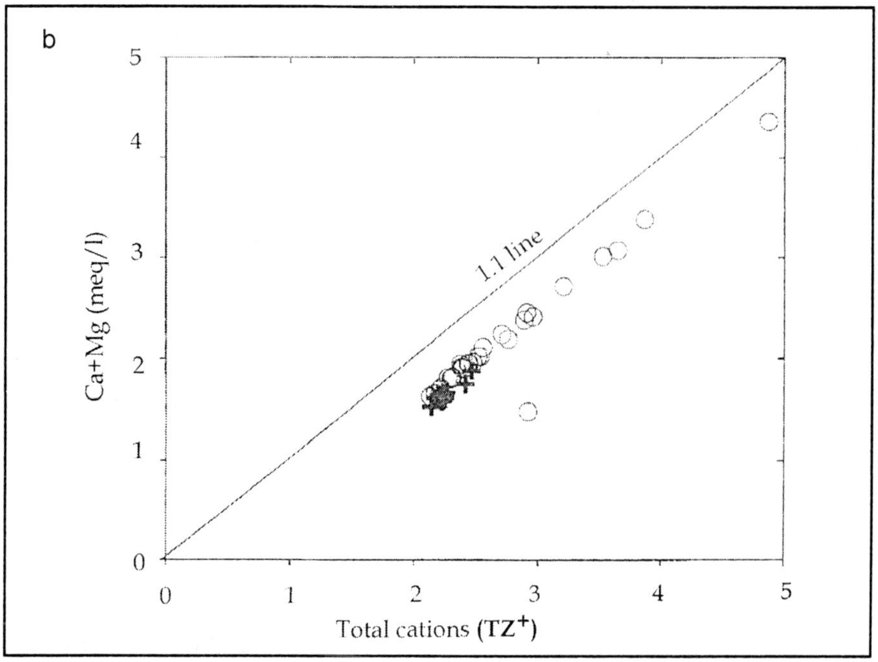

Fig. 5.6 (b) In a plot of $(Ca^{2+} + Mg^{2+})$ vs Tz^+ (total cations) the samples of both the seasons fall below the equiline in a linear trend.

Surinsar Lake

Surinsar Lake is about 11.70 km N-W of Mansar in the same strike continuation in Jammu district. The lake area is about 0.28sq. km and the maximum depth is 23.20 meters in the centre of the lake. It is at an elevation of 600 m above msl and formed during Holocene period. The lake is situated at the crestal part of a WNW-NW and ESE-SE trending anticlinorium known as Surin-Mastgarh anticline. It is about 900 m in length and 500m in width and is confined within the limits of 75^0 5' 11" to 75^0 5' 12" E and longitude 32^0 40' 58" to 32^0 40' 59" N latitude. It is a closed basin lake since no stream or nallah is feeding it nor there is any visible channel draining out (Fig. 5.7). The catchment area on all sides is very limited, less than 1sq. km. The lithology of the catchment comprises lower Siwalik Group of rocks consisting of sandstone alternating with thin clay-shale beds trending NW-SE with steep dips on either sides. The Lake is primarily fed from precipitation, surface runoff and underground springs. The rainfall is about 80-100 cms/yr and is influenced by SW-Monsoon. The population of the area is sparse hence the anthropogenic interference is less except along the banks.

Fig. 5.7. A panoramic view of Surinsar Lake showing a island in the middle of the lake and growth of vegetation and weeds near the bank indicating anthropogenic pollution.

The water chemistry of Surinsar Lake is similar to Mansar. The pH ranges from 8.38 to 8.56 with an average of 8.44 in the spring season. The E.C is 196 μs/cm and TDS ranges between 110-120 mg/l indicating moderate ionic strength of this lake because increase in ionic strength enhances electrical conductivity (EC) in the lake water (Wetzel, 1983) (Table 5.5). A little higher pH of Surinsar water compared to Mansar during this season may indicate relatively higher photosynthesis removing CO_2 from water consequently raising pH. The Chemistry of water shows that Ca^{+2} and Mg^{+2} together accounts for 84 to 86 equivalent percentage of the total cation budget and Ca^{+2} alone accounts for 71 equivalent% whereas $HCO_3^- + SO_4^{-2}$ constitute 86 to 88 equivalent% of total anions and HCO_3^- alone accounts for 86 equivalent%. Therefore, the percentage of Ca^{+2} + Mg^{+2} as cations and HCO_3^- as anions are almost equal but HCO_3^- over Ca^{+2} is in excess (Table 5.6). Hence, the Chemical composition of Surinsar Lake water is Ca^{+2} – HCO_3^- type. This may reflect that the water is influenced by calcium carbonate available in sedimentary sequence, though the rocks of catchment area are predominantly sandstone, and shale interlayered with limestone. Calcarenite (Calcite) occurs in much subordinate amounts in the matrix of sandstone. Nevertheless, the tendency of strong dissolution of calcite releases much higher percentage of calcium in water (Berner and Berner, 1987). The plot of TDS vs Na/(Na + Ca) suggest rock weathering as the dominant source of chemical composition of water (Fig.5.8).

The average (Ca + Mg)/HCO_3 equivalent ratio 1.64 and (Ca + Mg)/Tz^{2+} ratio 0.85 may indicate both silicate and carbonate dissolution to yield (Ca + Mg) to the water but the former will require alkaline earth silicate terrain to justify its primary source. But the regional geology of the area does not reflect this. The stoichiometry of carbonate weathering reaction demands that carbonate derived Ca and Mg should be equal to the carbonate derived bicarbonate (Singh and Hasnain, 1998). Since this ratio in Surinsar water is 0.88, this indicates that HCO_3 is in excess to Ca + Mg suggesting contribution from other sources than carbonates. The (Na+K) /Tz^+ ratio is an index to assess the contribution of cations via silicate weathering (Stallard and Edmond, 1983, 1987). This ratio for Surinsar Lake is 0.15 and Ca + Mg : Na + K ratio is 5.83 meq/l indicating that calcareous sandstone and shales of the lower Siwalik Group are the major source of lake water chemistry.

The measured partial pressure ($PCO_2^{-3.7}$) of water is close to atmospheric $PCO_2^{-3.5}$ indicating that the water is in equilibrium with atmospheric carbondioxide. However, in Mansar Lake it is higher ($PCO_2^{-2.4}$) having capability to dissolve more Ca^{+2} which in summer season due to enhanced photosynthesis precipitates Ca^{+2} resulting in its depletion in water. The water of Surinsar Lake is supersaturated with respect to both calcite and dolomite (Fig.5.9) but the Mansar Lake water is undersaturated in summer season due to stratification of lake dissolving more CO_2 in water.

Table 5.5: Chemical Characteristics of the Surinsar Lake

Sample No.	pH	E.C. µs/m	D.O.	CO_3^{-2}	HCO_3^-	Cl^-	SO_4^{-2} Mg/l	PO_4^{-3}	SiO_2	Na^+	K^+	Ca^{+2}	Mg^{+2}	T.H.	TDS
Sur-1	8.49	199	10.1	nil	122.0	11.36	0.5	0.026	1.49	10.79	2.76	60	5	170.46	110
Sur-2	8.46	198	8.8	nil	122.0	9.94	0.5	0.015	1.39	10.79	2.76	57	5	162.97	120
Sur-3	8.51	196	7.9	nil	122.0	11.36	0.5	0.000	1.55	11.47	2.42	57	5	162.97	110
Sur-4	8.39	199	8.2	nil	122.0	11.36	0.5	0.035	1.55	12.14	2.42	54	6	159.59	120
Sur-5	8.43	196	8.8	nil	146.4	11.36	0.5	0.000	1.29	11.47	2.76	60	5	170.46	110
Sur-6	8.50	195	8.0	nil	146.4	11.36	0.5	0.029	1.38	12.14	2.76	57	8	175.31	110
Sur-7	8.52	196	8.2	nil	122.0	11.36	0.5	0.029	1.38	11.47	2.42	54	5	155.47	110
Sur-8	8.41	193	8.1	nil	122.0	11.36	0.5	0.018	1.24	12.14	2.76	57	5	162.97	110
Sur-9	8.56	196	11.5	nil	122.0	11.36	0.5	0.013	1.41	12.14	2.42	54	9	171.94	110
Sur-10	8.38	194	9.5	nil	122.0	11.36	0.5	0.021	1.71	12.14	2.76	57	5	162.97	120
Sur-11	8.43	196	7.9	nil	122.0	11.36	0.5	0.026	1.29	11.80	2.76	57	11	187.66	110
Sur-12	8.29	198	7.8	nil	122.0	11.36	0.5	0.013	1.69	12.14	2.76	57	9	179.43	120
Sur-13	8.43	198	8.0	nil	122.0	11.36	0.5	0.015	1.60	12.14	2.42	54	8	167.82	120

Table 5.6: Percentage of equivalent per million of the constituents in the Surinsar Lake.

Sample No.	Cations			Anions		
	Ca	Mg	(Na+K)	S04	Cl	$(HCO_3^- + CO_3^-)$
Sur-1	75.76	10.61	13.63	0.43	13.73	85.84
Sur-2	74.80	11.02	14.17	0.43	12.23	87.33
Sur-3	74.41	10.97	14.62	0.43	13.73	85.84
Sur-4	71.24	13.19	15.57	0.43	13.73	85.84
Sur-5	75.19	10.53	15.28	0.43	13.73	87.91
Sur-6	69.17	16.26	14.56	0.43	11.72	85.84
Sur-7	73.37	11.41	15.22	0.43	13.73	85.84
Sur-8	73.64	10.85	15.50	0.43	13.73	85.84
Sur-9	66.83	18.56	14.60	0.43	13.73	85.84
Sur-10	73.64	10.85	15.50	0.43	13.73	85.84
Sur-11	65.52	21.15	13.33	0.43	13.73	85.84
Sur-12	67.86	17.86	14.28	0.43	13.73	85.84
Sur-13	68.18	16.92	14.90	0.43	13.73	85.84

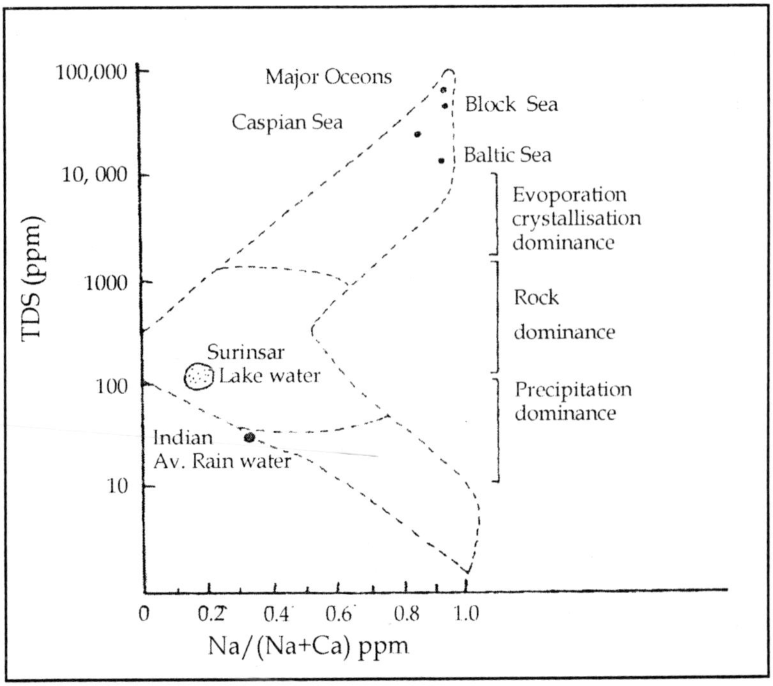

Fig. 5.8. Plot of TDS vs $N^{a+}/(N^{a+} + Ca^{2+})$ of Surinsar Lake showing predominance of rock weathering in the area.

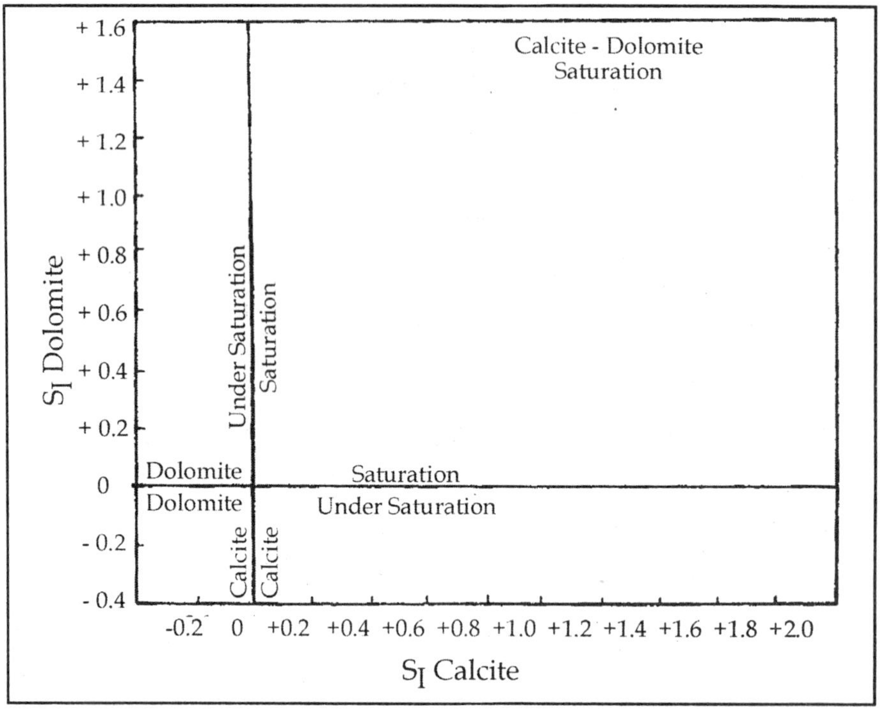

Fig. 5.9. Saturation indices (SI) plot of Surinsar Lake indicating that the water is supersaturated with respect to both calcite and dolomite.

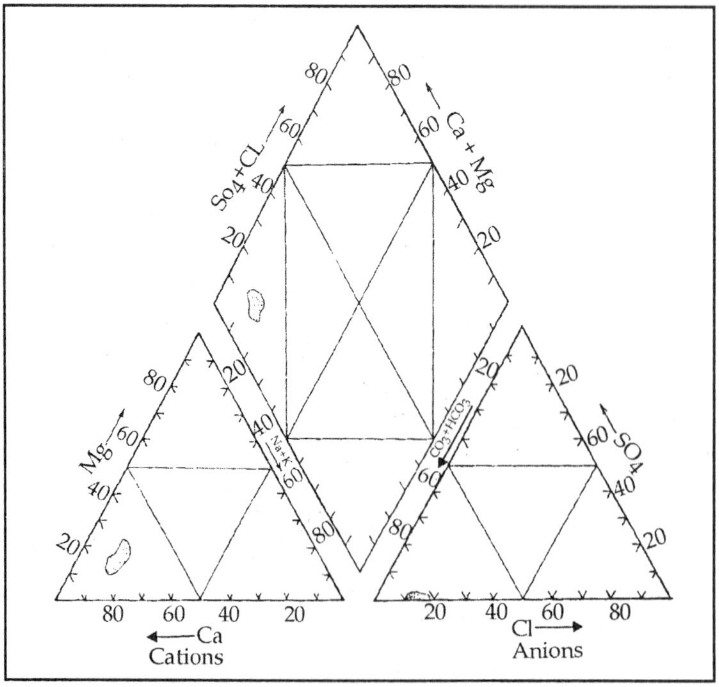

Fig. 5.10. Hydrological facies diagram of Surinsar Lake showing that the water is Ca^{2+} - Mg^{2+} - HCO_3 - type.

The hydrological facies which denote the diagnostic chemical character of water in hydrological system, plotted on trilinear diagram (Fig.5.10) for Surinsar Lake show Ca-Mg-HCO_3 type facies.

Lakes of Spiti valley (Higher Himalaya)

The high altitude lakes of North-West Himalaya occupy an intermediate position between Ladakh lakes exposed to cold desert conditions (Hutchinson 1937) and Central Himalayan lakes which receive abundant summer monsoon rains (Loffler 1969). In contrast, the lakes of western Himalaya receive scantly summer rains but subjected to long wet winter season. The high altitude lakes are mostly oligotrophic and drain through snow melt and precipitation. The mid and low elevation lakes and wet lands receive water from local rains through streams and nullahs. Very little limnological data are available on these highly diverse ecosystems because of difficult approach and inhospitable terrain.

Dankar, Thinam and Gete lakes of Spiti valley in the western Himalaya, have been investigated which elucidate major control of water chemistry by the respective basin catchment lithology. As the area is sparcely populated, human interference is presumably negligible and the lakes reflect their pristine character.

Dankar is the largest of the three lakes having a depth of > 3m and the water temperature ranging between 16.5 °C to 17.6°C in the month of October and that of the atmosphere 18.4°C. The lakes are situated at an altitude of 4200 m above msl. The catchment of Dankar lake comprises of Lilang Group of rocks (Triassic) comprising of black limestone intercalated with shale, dolomite and sandstone which attains enormous thickness at places. It is marine and fossiliferous. The lake Thinam is situated on the Spiti Formation (Jurassic) whereas Gete is at the contact of Spiti and Lilang Group of rocks (Fig.5.11). The Spiti Formation comprises of black shales, fine grained, sandstone and siltstone. The black shales contain abundant calcareous concretions enclosing fossils. It sporadically shows gypsum scales. There are also a number of thin limestone and calcareous sandstone bands within the black shales.

The Dankar and Thinam Lakes are alkaline with pH ranging from 8.32 to 10.19 and 9.85 to 10.45, respectively. Gete, on the other hand, is almost neutral to very mildly alkaline (pH 7.30-8.04). The Ca^{2+} and Mg^{2+} on average are 37.7 mg/l and 51.47 mg/l,HCO_3^- 58.96 mg/l and carbonate alkalinity 0.98 meq/l in Dankar water (Das and Dhiman, 2003). The lake is enriched in Mg^{2+} compared to Ca^{2+} (Table 5.7). As the lake catchment comprises of limestone, dolomite and shale of Lilang Formation, dissolution of carbonate

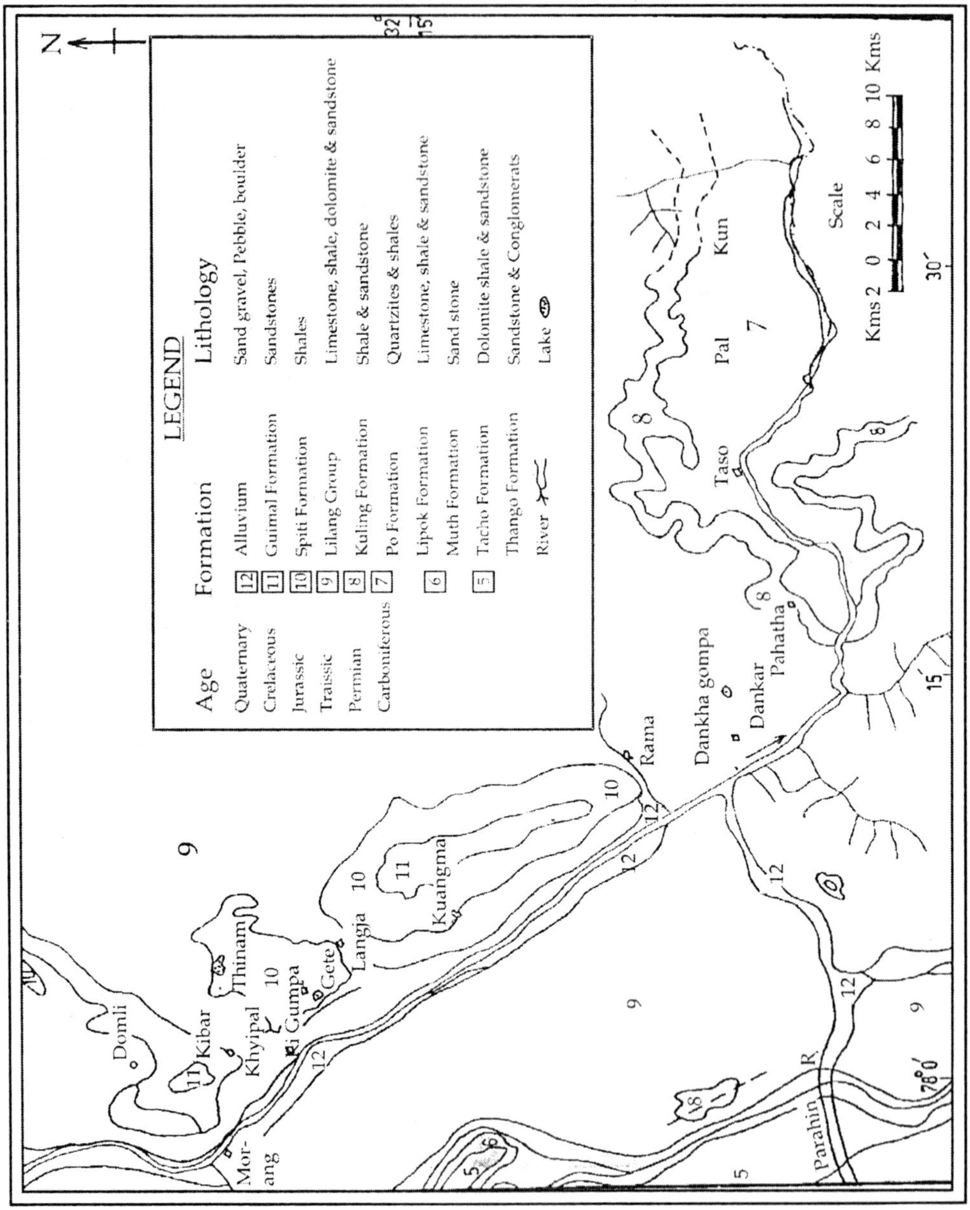

Fig. 5. 11. Generalized map of spiti showing lithology and location of Dankar, Thinam and Gete Lakes.

minerals are expected during weathering, which enhances E.C and TDS of the solution (Fig.5.12). The source of protons in the rapid weathering of carbonates in this case is carbonic acid and as the equivalent ratio of dissolved Ca: HCO_3 in the water is 1:2, it is resulting from calcite weathering.

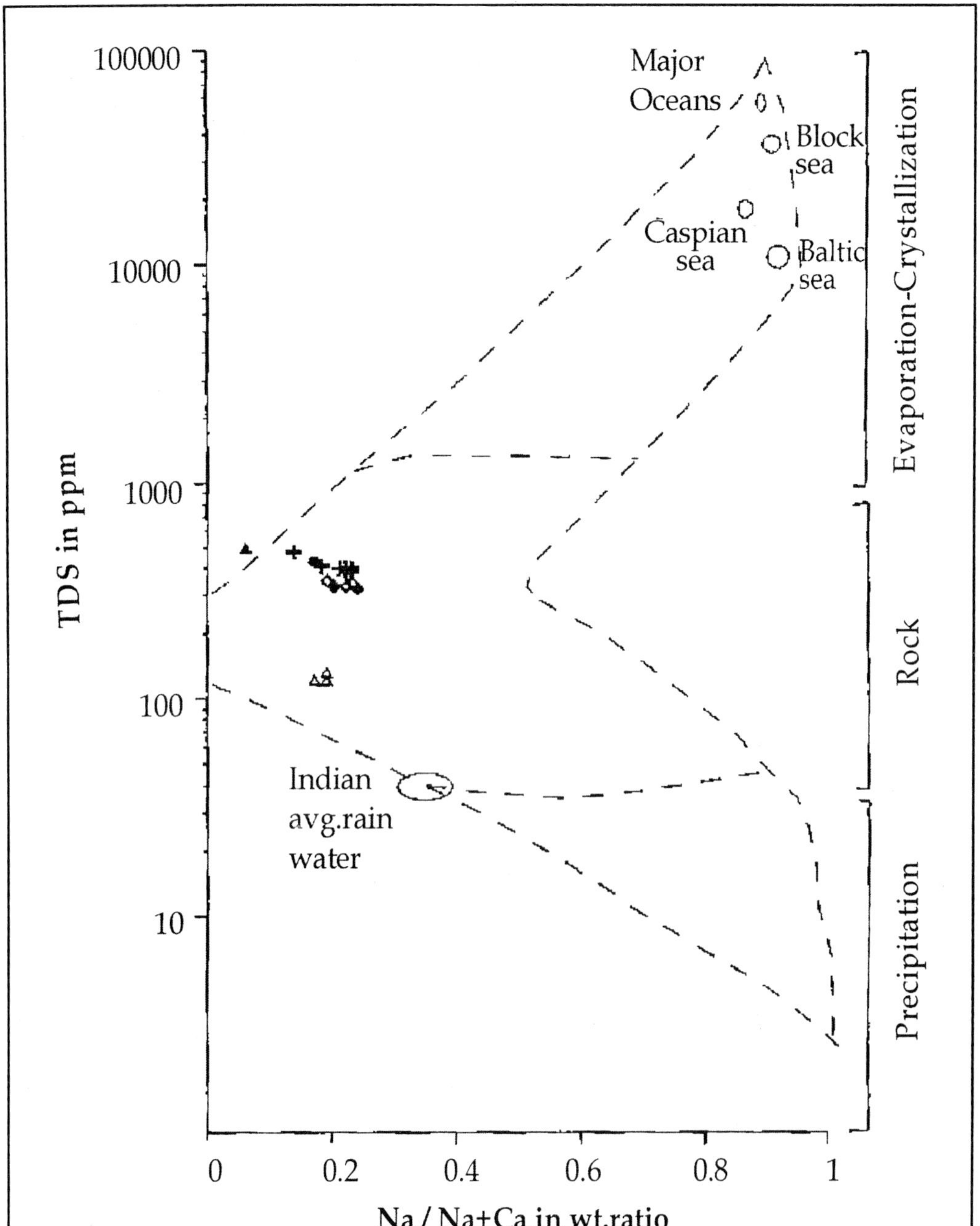

Fig. 5.12. TDS vs Na/Na + Ca plot of water chemistry of Spiti Lakes showing rock dominant ion chemistry of the lake.

Table 5.7: Chemical Characteristics of Higher Himalayan Lakes

Location No.	pH	EC µs/cm	DO	TDS	CO_3^{2-}	HCO_3^-	SO_4^{2-}	Cl^-	SiO_2 mg/l	PO_4^{3-} (µg/l)	Ca^{+2}	Mg^{+2}	Na^+	K^+	T.H.
LAKE DANKAR															
D_1	10.07	632	9.6	400	6.0	48.8	245.0	8.52	1.27	2.9	32.4	51.0	9.89	0.59	290.85
D_2	10.12	630	9.2	390	3.0	36.6	263.0	8.52	1.22	2.9	32.4	52.2	9.58	0.88	295.79
D_3	10.09	628	7.8	390	3.0	48.8	263.0	8.52	1.20	5.8	34.4	51.2	9.89	1.18	296.67
D_4	10.19	633	7.0	390	6.0	48.8	260.0	8.52	1.22	7.8	34.8	52.5	10.50	0.59	303.02
D_5	10.05	631	6.4	390	3.0	48.8	260.0	8.52	1.13	5.8	35.4	50.7	10.50	0.59	297.11
D_6	9.98	637	6.9	400	6.0	48.8	273.0	9.94	1.58	6.8	34.7	50.7	9.89	0.59	295.36
D_7	9.96	647	5.1	400	6.0	61.0	237.0	9.94	1.60	7.8	36.0	51.2	9.58	0.59	300.66
D_8	9.01	662	5.6	410	9.0	79.3	245.0	8.52	5.66	9.7	41.2	52.8	9.27	1.18	320.24
D_9	8.32	760	10.3	480	12.0	109.8	245.0	8.52	4.98	5.8	58.2	51.0	9.27	1.47	355.30
LAKE THINAM															
K_1	9.85	206	6.4	120	0.0	61.0	64.0	7.10	1.62	4.8	30.0	8.9	6.18	0.59	111.57
K_2	10.45	217	8.8	130	0.0	67.1	55.0	7.10	2.11	4.8	29.4	7.6	6.79	0.69	104.72
K_3	10.29	206	7.5	120	3.0	61.0	55.0	7.10	2.82	6.8	25.5	10.2	6.18	0.59	105.68
K-spring	7.91	768	17.8	490	6.0	85.4	255.0	9.23	5.80	4.8	89.6	27.7	5.56	1.77	337.83
LAKE GETE															
GT-1	8.71	520	6.5	330	6.0	61.0	214.0	7.81	4.51	7.80	68.7	18.1	19.78	2.07	246.11
GT-2	7.30	541	4.8	340	6.0	61.0	215.0	7.10	3.73	0.9	64.7	20.2	19.78	1.18	244.76
GT-3	8.04	531	7.1	330	3.0	61.0	200.0	8.52	3.60	4.8	64.8	17.0	18.23	1.77	231.84
GT-4	7.64	497	4.9	350	3.0	54.9	226.0	8.52	3.27	14.6	74.4	16.3	17.92	1.77	252.94
GT-spring	7.83	530	6.3	330	6.0	97.6	172.0	7.10	3.31	3.9	71.4	14.0	18.23	1.77	235.98
LAKE GETE(A)															
GIA	7.62	511	5.7	320	6.0	85.4	141.0	7.10	2.20	4.8	58.8	11.0	18.55	1.77	192.16
GIIA	7.34	675	11.9	430	12.0	146.2	168.0	7.10	9.07	4.8	82.5	18.0	16.69	2.36	280.17

$$CO_2 + H_2O + CaCO_3 \rightarrow Ca_2^+ + 2HCO_3^-$$

However, weathering of carbonate minerals does not show a balance between $Ca^{2+} + Mg^{2+}$ and HCO_3^- on an equivalent basis and indicate deficiency of HCO_3^- in water (Fig.5.13a).

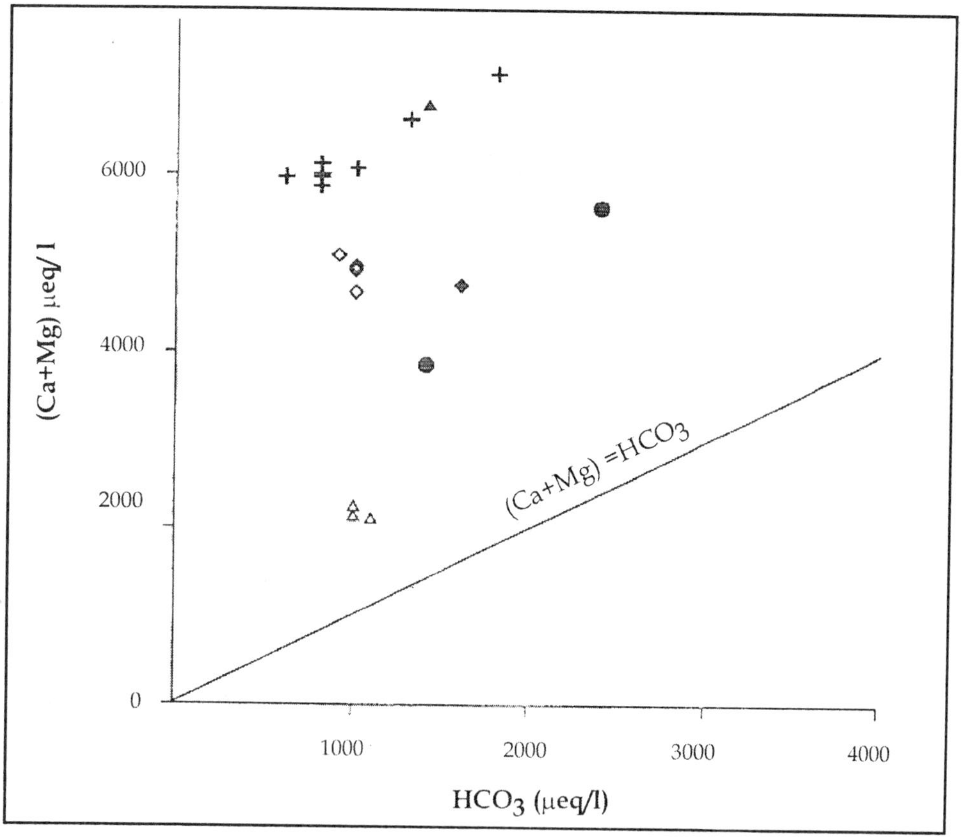

Fig. 5.13 (a). Ca/Mg vs HCO_3 plot of Dankar, Thinam and Gete Lakes water showing scattering of points above uniline.

This has been balanced by other anions such as SO_4^{2-} in this case derived from black pyrite shales due to reduction process during weathering (Fig.5.13b). The major contribution of (Ca+ Mg) to the total cations (Tz^+), high (Ca +Mg)/ (Na+K) ratio (>6) and very low (Na +K)/ Tz^+ (0.07) indicate that limestone intercalated with black shale and dolomite of Lilang Group, are the major source of lake water chemistry.

For calcite

$$H_2CO_3 + CaCO_3 \rightarrow Ca_2^+ + 2HCO_3^-$$

and for dolomite

$$2H_2CO_3 + CaMg\ (CO3)_2 \rightarrow Ca_2^+ + Mg_2^+ + 4HCO_3^-$$

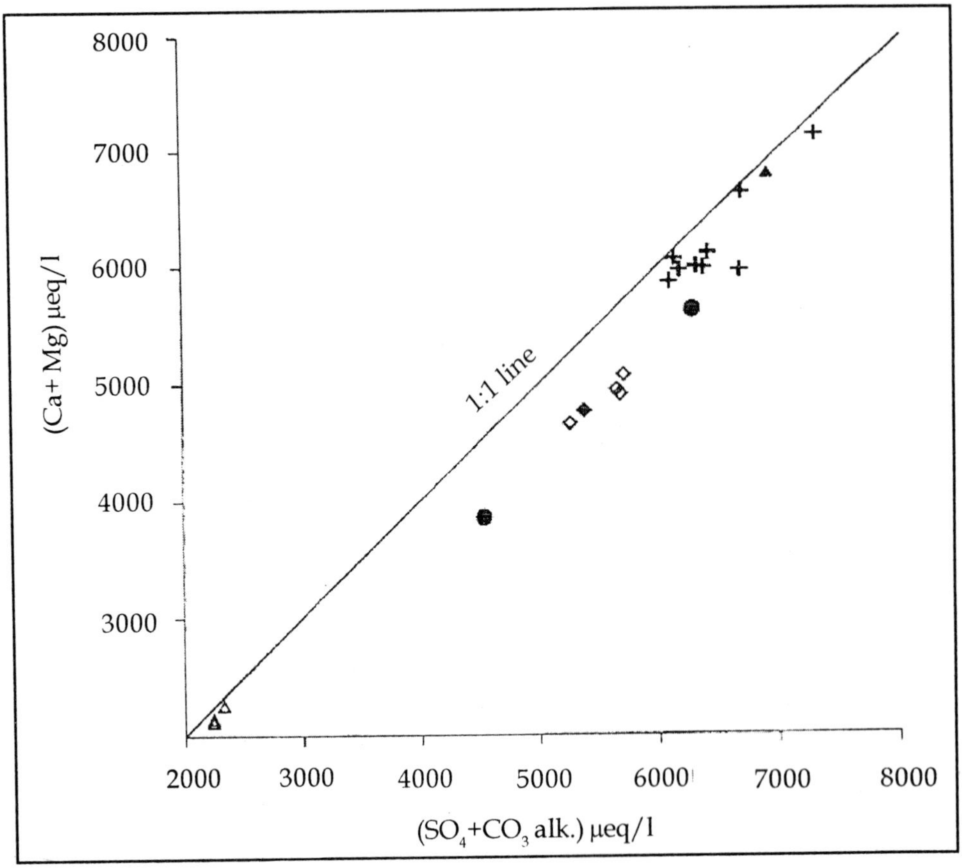

Fig. 5.13 (b). In Ca/Mg vs (SO_4 + CO_3 alk) plot of Spiti Lakes, points fall below 1:1 line indicating contribution of alkalies to balance the charge.

In Thinam Lake Ca^{2+} is 28.3 and Mg^{2+} 8.9 mg/l & HCO_3^- is 63.03 mg/l. The total alkalinity is 1.06 meq/l. The low Ca^{2+} and still lower Mg^{2+} content, but relatively higher HCO_3^- in this lake is due to photosynthesis during which CO_2 escaped making water supersaturated and consequent precipitation of calcite. This has elevated pH of the water. As the PCO_2 in the lake water is lower relative to the ambient $PCO_2^{10-3.5}$, dissolution of calcite is restrained, depleting Ca^{2+} and Mg^{2+} and other solutes in the lake (Table 5.7). However, as the dissolution of calcite is faster than silicate minerals in fresh water, except when water is more acidic, its concentration is higher than other cations in the lake even though carbonate rocks are not the predominant lithology of Spiti Formation as mentioned earlier. However, (Ca+Mg): Tz^+ ratio (0.87) may indicate both carbonate and silicate weathering as source of ions. The presence of bicarbonate excess over (Ca+Mg) in waters and the relatively high (Na +K)/ Tz^+ ratio (0.12) seems to indicate that the supply of cations to the water by silicate weathering is also likely.

The investigation of groundwater (spring) shows low pH of 7.91 and higher $PCO_2^{10.35}$ indicating more CO_2 in water, increased salinity (HCO_3^-) and high concentration of Ca^{2+} and other solutes in the water.

On equivalent basis, water of Thinam Lake shows enrichment of major cations $Ca^{2+}$and Mg^{2+} and major anions are HCO_3^- and SO_4^{2-}, which are almost equal in concentration. The $Ca^{2+} + Mg^{2+}$: HCO_3^- equivalent ratio (2.09) suggests weathering of carbonate rocks releasing these cations and anions to the water. A plot of ($Ca^{2+} + Mg^{2+}$) vs HCO_3^- (Fig.5.13a) shows that the samples lie above the 1:1 line indicating that ($Ca^{2+} + Mg^{2+}$) are in excess to HCO_3^- in these waters and is balanced by (SO_4^{2-} +CO_3 alkalinity) (Fig.5.13b). Though the major contribution is of ($Ca^{2+} + Mg^{2+}$) to total cations (Tz^+), but relatively low Ca^{2+} + Mg^{2+}: Na^+ K^+ ratio (3.6) and higher ($Na^+ + K^+$)/ Tz^+ ratio (0.12) than Dankar Lake indicate contribution of silicate rocks of the catchment in the water chemistry of this lake.

Lake Gete, in contrast to the other two lakes, shows lower pH, higher conductivity, TDS and higher HCO_3^-, SO_4^{2-} and considerably higher Ca^{2+}, Na^+ and SiO_2 concentration. The higher HCO_3^- and Ca^{2+} justify low pH and high PCO^2 of the lake water which tends to dissolve more of Ca^{2+} consequently elevating Ca^{2+} and HCO_3^- in the solution. Similarly, higher sulphate content in the lake water is derived from black pyritiferrous shale and gypsum streaks in the lithology of Spiti shales and Lilang limestone.

Gypsum weathers faster than pyrite and goes into solution faster releasing $SO4^{2-}$ ions than in pyrite. This lake is not very productive and the photosynthesis is not very significant. The lake is, therefore, undersaturated to saturated with respect to calcite and dolomite as is evident from saturation indices plots (Fig.5.14). Like the other two lakes, water of Gete is also in excess of (Ca^{2+} + Mg^{2+}) over HCO_3^- and the deficiency of HCO_3^- is balanced by $SO4^{2-}$ + CO_3 alkalinity. However, all cations are not derived from carbonate weathering. In Gete Lake, there is a definite excess of sodium over chloride and the former is ~ 4 times in excess of chloride on equivalent basis. This excess (Na*; Stallard and Edmond, 1983) is attributed to silicate weathering. The (Na* +K)/Tz^+ ratio is an index to assess the contribution of cations via silicate weathering (Stallard and Edmond, 1983, 1987).

This ratio for lake Gete is 0.15 and there is an excess of kaolinite in the sediments, relative to other two lakes, suggesting contribution of cations to these waters by silicate weathering. The chemistry of lake waters of these lakes demonstrates variations controlled by the catchment lithology.

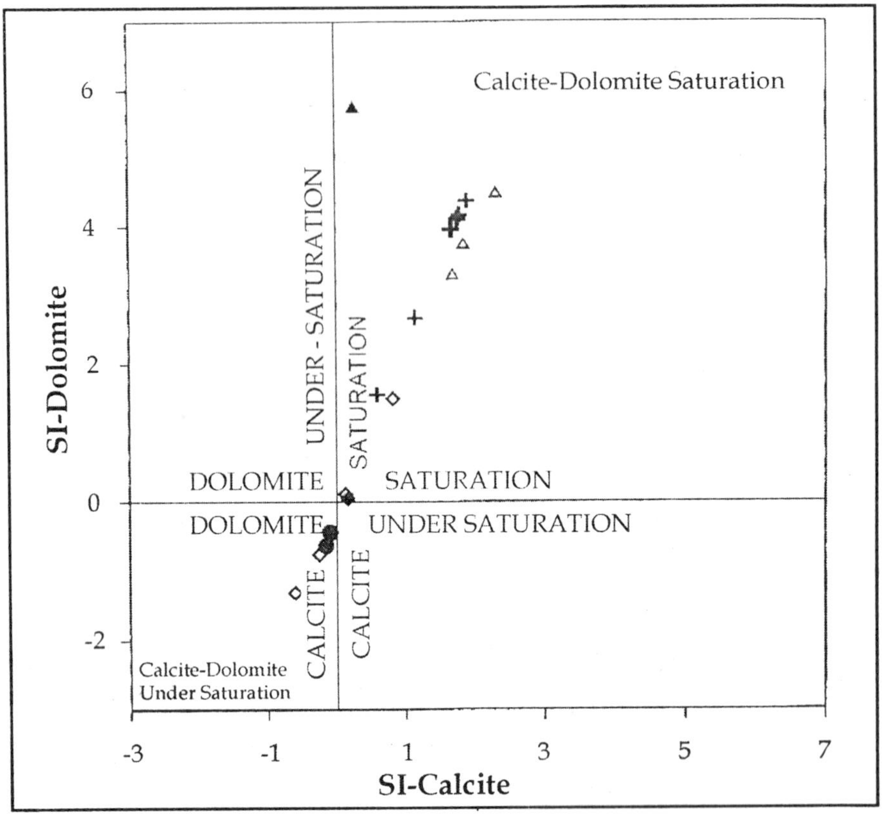

Fig. 5.14. Saturation Indices plot showing that Dankar (+) and Thinam (Δ) water is supersaturated with respect to Calcite and Dolomite but Gete Lake (•) water is undersaturated.

Lakes of Himachal Pradesh

Renuka Lake

The oval-shaped Renuka Lake is located at an elevation of 620 m above msl in Sirmaur district of Himachal Pradesh. The lake is 1050 m is length and 204 m in width having maximum depth of about 13 m and the catchment area is ~ 254.3 ha. The annual rainfall is between 150 and 200 cm. The lake follows a riparian course between steep hill slopes on both the sides with thick forest and is considered to occupy abandoned course of Giri River (Fig. 5.15a,b). There are inflowing and out flowing channels in Renuka Lake and the water is supplied through precipitation and springs within the basin. The macrophytic vegetation covers an area of 39,969 m^2 in the lake. The human interference, such as deforestation, construction of road, human settlements near the shore helps more silt flow into the lake, particularly during rainy season. The rate of sedimentation is 3.3 mm/yr measured by Pb^{210} isotope method (Das and Kaur, 2001).This is imposing threat to lake environment.

Fig. 5.15 (a). A panoramic view of Renuka Lake

Fig. 5.15 (b) Photo Shows hill slop with dense vegetation and settlements along the bank of Renuka Lake.

The Catchment rocks of Renuka Lake comprises Infra-Krol dark shales, slates, black carbonaceous shales or slates, and slaty quartzite, purple-red shales with intercalation of green shales, thin dolomite and cherty limestones of the Krol Group, Blaini boulder bed with limestone and well bedded grayish white limestone and shales of the Krol Group (neo-Proterozoic-Proterozoic) (Fig.2.4).

The Renuka water is dominated by calcium and magnesium ions and the former contributes 38.57 to 48.93% and latter 44.71 to 54.16% of the total cation budget. The bicarbonate contribute 36.97 to 51.90% and sulphate (SO_4^{2-}), almost in the same proportion, 43.25 to 51.91%.

The contribution of (Na +K) is between 4 to 7% and chloride is < 6% to the major ion budget of the lake (Tables 5.8 and 5.9). The high Ca^{2+}+ Mg^{2+} and HCO_3^- in water indicate dissolution of limestone in the lake basin. This is also reflected from (Ca^{2+} +Mg^{2+}): HCO_3^- equivalent ratio of ± 2.55. However, presence of bicarbonate in excess of (Ca + Mg) in water may indicate contribution from silicate rock weathering as well.

Since the limestone ($CaCO_3$) weathers much more rapidly than silicate minerals in sedimentary sandstone and shales, releasing Ca^{2+} and Mg^{2+} as cations and $HCO3^-$ as anions in solution, the Renuka waters are enriched in these constituents (Berner and Berner, 1987).

Since the rocks exposed to weathering are in contact with dissolved oxygen (rainwater), pyrite being chemically unstable and rapidly undergo oxidative decomposition. This results in the production of sulphuric acid, which brings about further weathering of silicate and carbonate minerals. The following reactions take place:

$$4FeS_2 + 15\,O_2 + 8\,H_2O \rightarrow 2Fe_2O_3 + 8\,H_2SO_4$$

$$H_2SO_4 \rightarrow 2H^+ + SO_4^{2-}$$ .... (Berner and Berner, 1987)

It is reported that if sulphuric acid is the weathering agent, then the Ca: SO_4 ratio in the waters would be 1:1 for calcite and 1:2 for dolomite. This ratio in Renuka water is little less than one (1:0.92) indicating a possibility of sulphuric acid replacing carbonic acid as a source of proton for rock weathering :

$$H_2SO_4 + 2CaCO_3 \rightarrow Ca_2+ + 2HCO_3- + SO4^{2-}$$

In this case, HCO_3^- arises only from the carbon in carbonate minerals and none comes from soil CO_2 (Berner & Berner, 1987).

Table 5.8: Water chemistry of Renuka Lake

Sample No.	pH	EC μs/cm	DO	TDS	ORP (mV)	CO^{2-}_3	HCO^-_3	SO_4^{2-}	Cl^-	PO^{3-}_4 (μg/l)	SiO_2	Ca^{2+}	Mg^{2+}	Na^+	K^+	Total Hardness
Renuka Lake																
RN-1	8.35	0.554	7.9	360	231	12	109.8	130	36.92	13.5	7.20	54.00	37.6	9.44	29.73	19.63
RN-2	8.47	0.548	8.2	360	234	12	158.6	150	8.52	3.5	6.10	57.50	37.2	8.09	1.38	19.93
RN-3	8.52	0.547	8.1	360	232	18	134.2	130	9.94	5.0	6.10	44.50	36.4	7.42	1.04	18.02
RN-4	8.42	0.549	7.0	360	235	18	134.2	130	9.94	7.0	6.60	46.00	37.6	7.42	1.04	18.63
RN-5	8.52	0.548	6.7	360	234	18	146.4	152	11.36	8.5	6.20	63.50	36.9	8.09	1.04	20.56
RN-6	8.50	0.550	8.4	360	233	12	158.6	170	9.94	1.5	6.20	66.50	36.4	8.09	1.04	20.77
RN-7	8.51	0.549	8.1	360	236	18	134.2	160	11.36	13.5	6.25	59.50	37.6	7.75	0.69	20.31
RN-8	8.55	0.552	8.6	360	237	12	146.4	190	11.36	15.5	6.10	63.00	35.8	7.42	1.04	20.14
RN-9	8.21	0.552	9.1	360	239	18	146.4	135	11.36	5.5	6.25	55.50	37.2	8.09	0.69	19.68
RN-10	8.40	0.550	8.6	360	237	18	146.4	120	9.94	1.5	6.24	48.00	36.4	8.09	0.69	18.47
RN-11	8.59	0.547	8.9	360	235	24	134.2	155	9.94	3.5	6.10	59.50	37.4	7.42	0.69	20.26
RN-12	8.55	0.545	9.9	360	236	18	146.4	150	11.36	3.5	5.88	57.50	38.7	7.08	0.69	20.43
RN-13	8.40	0.544	10.0	360	238	12	146.4	150	11.36	12.0	5.70	58.00	37.2	6.77	0.35	20.01
RN-14	8.40	0.545	8.0	360	241	18	134.2	160	9.94	8.5	6.20	59.00	38.0	7.42	0.69	20.42
RN-15	8.51	0.541	9.2	360	236	12	158.6	140	9.94	5.0	6.00	58.00	36.6	6.07	0.69	19.80
RN-16	8.42	0.540	9.6	360	236	24	134.2	140	11.36	1.5	6.10	64.00	36.8	6.07	0.69	20.63
RN-17	8.40	0.531	9.5	350	238	24	122.0	135	11.36	5.0	6.10	51.50	36.6	6.41	0.35	18.98
RN-18	8.54	0.523	8.6	350	240	24	122.0	140	9.94	8.5	6.20	60.50	39.4	8.77	0.69	21.05
RN-19	8.38	0.524	8.4	350	246	0	150	150	12.78	3.5	6.00	56.40	37.2	10.14	1.04	19.81
RN-20	8.35	0.525	8.5	340	240	24	140	140	12.78	10.0	6.24	54.00	37.6	9.44	1.04	19.63
RN-21	8.35	0.523	8.1	330	237	24	140	140	9.94	5.0	6.08	56.00	37.2	9.44	0.69	19.76
RN-22	8.34	0.523	8.2	330	232	18	142	142	10.65	1.0	5.78	57.60	36.2	9.10	1.04	19.63
RN-23	8.34	0.525	7.5	330	229	12	131	131	10.65	7.0	5.70	52.80	38.4	8.77	0.35	19.77
RN-24	8.34	0.532	7.0	340	227	12	146	146	9.94	7.0	6.10	52.00	36.2	8.09	1.04	18.93
Paras Ram Tal Lake																
PRT-1	7.60	0.606	7.1	420	233	18	195.2	180	11.36	10.0	11.40	73.60	46.8	8.42	2.07	25.25
PRT-2	7.92	0.667	7.2	450	229	18	170.8	230	14.20	3.5	9.40	80.80	49.5	11.46	2.07	27.05
PRT-3	8.23	0.617	6.6	420	224	30	158.6	200	14.20	8.5	5.00	55.20	45.3	12.14	2.07	22.42

N.D. = Non Detectable

Table 5.9. Percentage of Equivalent Per Million of Constituent in Renuka Lake

	CATIONS			ANIONS		
Sample No.	**Ca^{2+}**	**Mg^{2+}**	**Na^{+}+K^{+}**	**SO_4^{2-}**	**Cl^{-}**	**HCO_3^{-}+CO_3^{2-}**
Renuka Lake						
RN-1	38.57	44.71	16.71	45.55	17.48	36.97
RN-2	45.20	48.82	5.98	49.06	3.77	47.17
RN-3	39.64	54.11	6.25	46.80	4.83	48.36
RN-4	39.79	54.15	6.05	46.80	4.83	48.36
RN-5	47.88	46.37	5.74	48.84	4.93	46.22
RN-6	49.33	45.02	5.65	51.91	4.10	43.99
RN-7	45.97	48.45	5.57	51.63	4.96	43.41
RN-8	48.61	45.99	5.40	55.93	4.52	39.55
RN-9	44.39	49.68	5.93	45.84	5.22	48.94
RN-10	41.38	52.24	6.38	43.25	4.84	51.90
RN-11	46.19	48.52	5.29	49.61	4.30	46.08
RN-12	44.70	50.15	5.14	48.45	4.97	46.58
RN-13	46.03	49.21	4.76	50.00	5.13	44.87
RN-14	45.66	49.07	5.26	51.95	4.37	43.68
RN-15	46.55	48.96	4.49	47.09	4.52	48.39
RN-16	48.85	46.87	4.27	46.79	5.13	48.08
RN-17	43.48	51.61	4.91	47.39	5.40	47.22
RN-18	45.07	48.95	5.97	48.67	4.67	46.67
RN-19	44.13	48.51	7.35	52.52	6.06	41.41
RN-20	43.06	49.92	7.02	45.76	5.64	48.59
RN-21	44.23	48.97	6.79	46.35	4.44	49.21
RN-22	45.57	47.78	6.64	47.82	4.85	47.33
RN-23	42.37	51.36	6.26	48.92	5.38	45.70
RN-24	43.33	50.33	6.33	49.75	4.58	45.66
Paras Ram Tal Lake						
PRT-1	46.00	48.75	5.25	46.47	3.96	49.57
PRT-2	46.38	47.30	6.31	54.49	4.55	40.95
PRT-3	38.82	53.02	8.16	51.04	4.89	44.06

Since the HCO_3^- is in excess of Ca + Mg in waters, its contribution, in addition to carbonate minerals, may be expected. The pH of the lake water is alkaline (Table 5.8) and the activity of water draining rocks where sulphide oxidation is taking place depends on the content of sulphides and the presence of other minerals especially carbonates, which readily neutralizes acid. Therefore, sulphuric acid replacing carbonic acid as a source of proton for rock weathering in Renuka is unlikely. The sulphuric acid is neutralized by $CaCO_3$ by the following reaction:

$$4FeS_2+15O_2+8H_2O+16CaCO_3 \rightarrow 2Fe_2O_3+16Ca^{2+}+8\ SO_4^{2-} + 16\ HCO_3^-$$

This has yielded Ca^{2+} - SO_4^{2-} - HCO_3^- rich lake waters. Moreover, high Ca^{2+} + Mg^{2+}: Na^+ + K^+ ratio (8.09) and low $(Na^++K^+)/\ Tz^+$ (0.058) does not favour significant contribution of ions from silicate weathering. In view of these, excess of HCO_3^- over (Ca + Mg) may be assumed to be derived from soil CO_2:

Carbonic acid (H_2CO_2) is produced by the oxidation of organic matter by microbes to CO_2

$$CH_2O + O_2 \rightarrow CO_2 + H_2O$$

(organic matter)

This CO_2 then combines with water to form carbonic acid which further partly dissociates to H^+ and HCO_3^-.

$$CO_2 + H_2O \rightarrow H_2\ CO_3$$

$$H_2\ CO_3 \rightarrow H^+ + HCO_3^-$$ (Berner & Berner, 1987)

Plot of (Ca + Mg) vs HCO_3 shows that samples fall above 1:1 line and are scattered indicating that (Ca + Mg) is in excess to HCO_3^- (Fig.5.16a) and the deficiency of HCO_3^- is balanced by (SO_4^{2-}+ CO_3^- alk) and then the samples fall close to the 1:1 line (Fig.5.16b).

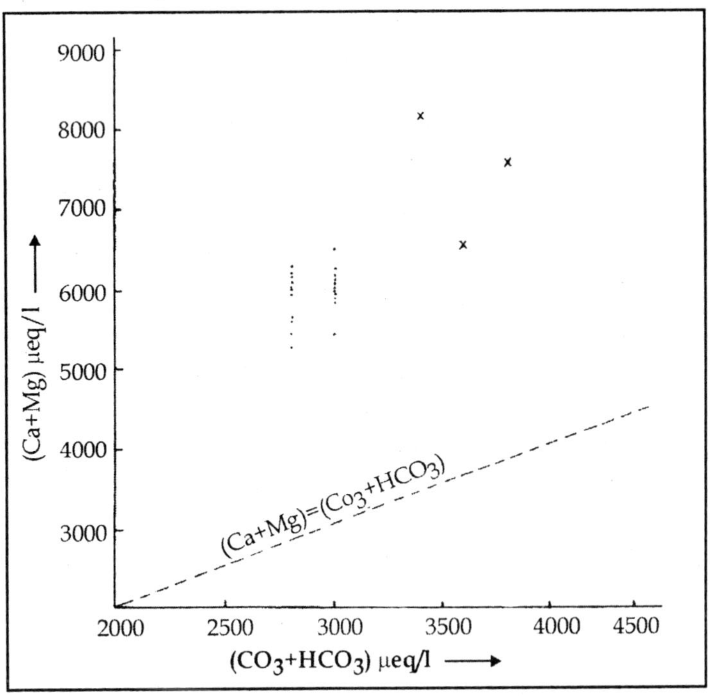

Fig. 5.16 (a). Ca/Mg vs CO_3+HCO_3 plot showing that the samples lie above 1:1 line and scattered indicating that (Ca +Mg) are in excess of HCO_3.

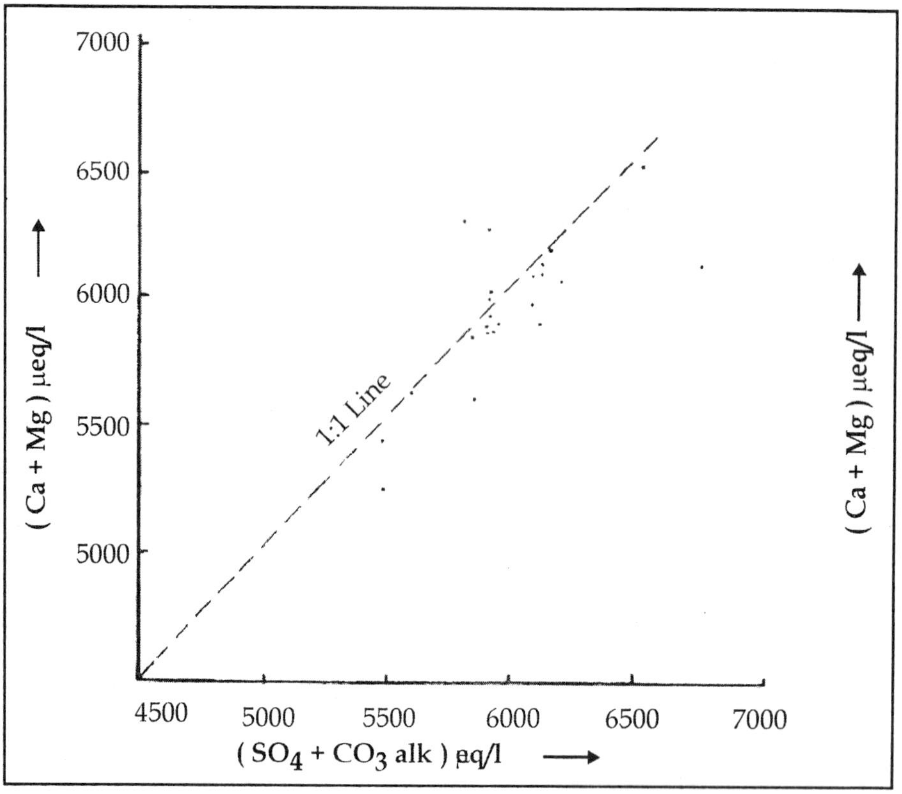

Fig. 5.16 (b) In a (Ca + Mg) vs (HCO_3 + CO_3alk + SO_4) plot points falls close to 1:1 line showing that the deficiency of HCO_3^- is met out.

Similarly, in a (Ca + Mg) vs Tz^+ plot the points lie in a linear trend close to 1:1 line (Fig.5.17). The very low K^+ content than Na^+ in the lake water may be due to its resistance to weathering as most of the ions are largely released during weathering from the constituent minerals of the rock and transported to the basin. Secondly, depletion of K^+ in the water may indicate its utilization in the formation of clay minerals- illite, which constitute between 52-90% of the clay mineral constituents of the sediments. Therefore, the water chemistry of this lake is primarily controlled by the weathering of carbonate rocks.

The plot of log $[Na^+]/[H^+]$ versus log $[H_4SiO_4]$, where brackets represent molar concentration in solution delineates the regions of stability for the various secondary minerals. Also, in this plot (Fig.5.18) Renuka water falls in kaolinite field indicating that kaolinite would be thermodynamically stable in the natural lake water environment. The secondary minerals identification has shown kaolinite in minor percentage after illite and chlorite. However, these clay minerals do not constitute in majority the bed sediments of the lake and thus are not in the state of equilibrium with the products of weathering. This is because the physical erosion has been stronger due to uplifting Himalaya, and thus in a weathering-limited condition. Eroded detritus (Quartz, illite rich) transported faster and deposited in the lake basin. Because transport of weathering material is rapid with high relief, chemical weathering in the soil is incomplete, resulting in a thin cover of bed rock in Renuka Lake.

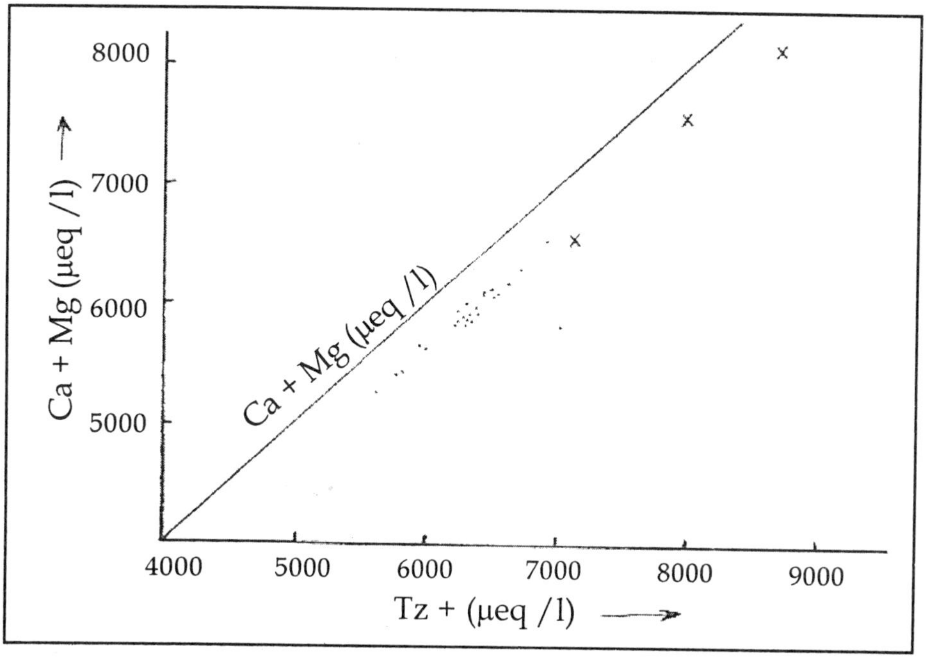

Fig. 5.17. In a plot of (Ca + Mg) vs. Tz⁺ points lie close to uniline in a linear fashion.

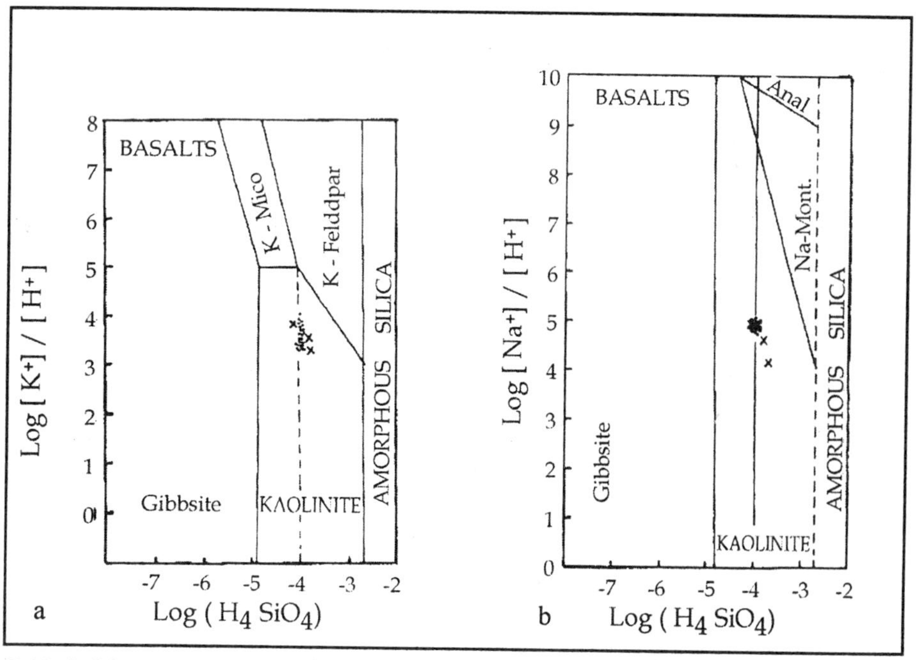

Fig. 5.18 (a,b) In a mineral equilibria diagram Renuka water plot in kaolinite field indicating that water is in equilibrium with kaolinite

Rewalsar Lake

The Rewalsar Lake is situated at an altitude of 1306 m above msl in district Mandi, Himachal Pradesh. It is a small~ 6 m deep, sub-tropical lake in sub-mountainous range of Western Himalaya. It has three seasonal inlets and one perennial outlet with a definite boundary bordered by thick patches of emergent macrophytes. The catchment area measures 173.12 hactares out of which nearly 53% is covered by agricultural fields. Rewalsar Lake is nearly square in shape (Fig.5.19 a & b) surrounded from the sides by hillslope and opens down hill into a nullah on the southeastern side. There is an undemarcated protected forest of 110 acres and type of vegetation is sub-tropical. Besides local population in adjoining town, the Hindu temples, Gurudwara an Buddhist monasteries attract large number of pilgrims throughout the year. This causes anthropogenic pollution in the lake. The annual precipitation is about 1690 mm per year and an average maximum temperature of 33°C are recorded. The minimum temperature touches freezing point and some times, snow fall occurs in the area. The lake primarily receives water from rainfall and seepage underground. The lake is known to have originated by tectonic activity and a strike fault running NE-SW resulted in the formation of depression in probably Holocene time.

Fig. 5.19 (a). A panoramic view of Rewalsar Lake showing configuration of lake, settlements and middle Siwalik Hills in the background.

Fig. 5.19 (b) Rewalsar Lake, Monastery, Temple and Gurudwara & middle Siwalik Hills along the bank of the lake.

The lithology of the catchment comprises of predominant micaceous, fine-grained, light coloured sand stone, with thin silt stone and shale with calcareous segregations and often ferruginous nodular red clays interlayers and belongs to middle Siwalik Group of rocks (Pliocene). The latter two are often thicker than the former indicating intermittent subsiding basinal conditions. On weathering, they yield clayey soil which is not very deep.

The bottom of the lake is composed of incompletely decomposed fibrous peat, zooplankton, and abundance of blue-green algae, macrophyte and the high presence of autocthonous suspended organic particulate matter, indicating eutrophic condition of this lake. Although there is no indication of complete stratification, however, a relatively stagnant condition and overturn towards the fall is indicated from the depletion of the nutrients, alkalinity and conductivity at the surface and a marked homogenous distribution of them during October (Singh *et al.*, 1986).

The Rewalsar Lake is neutral to mildly alkaline with low pH ranging from 6.95 to 7.24 due to higher dissolution of CO_2 resulting from increased respiration and organic decay in the basin (Table 5.10). This is in conformity with the higher partial pressure of carbondioxide (PCO_2 $^{10-2.09}$) indicating oxidation of organic matter and respiration in the basin. The relatively low pH of the samples along bank compared to the center, indicates strong anthropogenic activity causing respiration and organic decay. Consequently, higher dissolution of CO_2 occurs which increases PCO_2 and pH as compared to water in the center of the lake. The large macropytic phytoplankton population and algal growth in the lake tends to support the above observations. The carbonate alkalinity 1.90 meq/l is low. The SiO_2 concentration is extremely low ranging between 0.22 to 0.98 mg/l in the lake water except in two samples in which it is 1.04 and 2.14 mg/l. Contrary to this, values in groundwater is 10.89 mg/l which is much higher and close to the world average of 13 mg/l. The major source of silica is from the degradation of alumino-silicate minerals. But the solution of silica derived from various rocks is modified by the silicic acid adsorbed to inorganic particles or complexed organically. The silicate complexes with iron and aluminum hydroxides decreases solubility of silicates in sediments. The groundwater and surface water draining through volcanic rocks contain more silica than plutonic rocks and very small amount from sandstone and sedimentary rocks. The lowest amount of silica are found in water draining from carbonate rocks. Adsorption is the only significant mechanism of inorganic precipitation and it increases within pH 4 to 9 and above pH of 10, adsorption decreases sharply.

The low silica concentration in Rewalsar Lake is due to low pH causing greater adsorption and consequent depletion in water. Additionally, water drains from sandstone and sedimentary rocks of the catchment releases less silica. Silica is also utilized by diatoms in the synthesis of their frustules (tests) which greatly modifies the flux rates of silica in lakes and streams. Increasing light, and to lesser extent a rise in water temperatures, are major factors initiating the development of diatom populations from the small residual planktonic populations. The water samples were collected during early summer when the atmosphere was clear and sunny favourable to enhance photosynthesis. High temperature favoured the diatom bloom, hence silica uptake by these micro-organisms is no wonder, which has caused further depletion in silica budget of the lake. The high concentration of silica in groundwater is not uncommon, but a big contrast in the values of the two may support the process of silica depletion/removal in the lake. The diatom activity at shallow depth of the lake is more common and hence in groundwater such activities are not likely to occur. The high elemental concentration in groundwater may be due to many reasons, for example more of H_2CO_3, high PCO_2, and movement of water. More of

Table 5.10: Chemical composition of Rewalsar Lake Water (all parameters are in mg/l except EC,ORP& PO_4^{3-})

Location No.	pH	EC (μs/cm)	ORP (mv)	DO	TDS	CO_3^{2-}	HCO_3^-	SO_4^{2-}	Cl^-	SiO_2	PO_4^{3-} (μg/l)	Ca^{+2}	Mg^{+2}	Na^+	K^+	T.H
RW-1	7.06	339	214	7.9	250	18	73.2	20.00	9.94	0.85	26.00	34.80	4.1	8.96	3.62	103.80
RW-2	6.95	333	152	7.7	220	6	85.4	20.00	10.65	0.42	23.00	31.00	3.7	8.65	3.37	92.67
RW-3	7.02	334	187	7.8	250	12	91.5	10.00	12.07	0.96	41.00	33.50	4.2	8.96	3.11	100.97
RW-4	7.01	341	218	7.7	230	12	85.4	13.33	12.07	1.04	39.00	33.60	3.9	8.65	3.11	99.98
RW-5	7.08	348	214	7.7	230	15	91.5	13.33	12.78	0.24	42.00	35.40	4.2	8.96	3.11	105.72
RW-6	7.07	346	195	7.9	220	9	73.2	13.33	12.07	0.38	36.00	34.80	4.5	8.35	3.11	105.45
RW-7	7.24	333	199	8.0	220	12	85.4	6.66	11.36	0.40	9.00	31.30	3.7	8.03	3.11	93.42
RW-8	7.06	337	204	8.0	230	12	85.4	16.66	12.78	0.22	20.00	33.80	4.1	8.96	3.11	101.31
RW-9	7.01	333	208	7.9	220	12	97.6	13.33	12.07	0.26	24.00	35.10	4.4	8.65	3.11	105.79
RW-10	6.97	335	211	8.0	220	12	85.4	13.33	12.07	0.40	68.00	33.70	3.9	8.65	3.37	100.23
RW-12	7.14	332	213	8.1	210	15	85.4	13.33	11.36	0.76	13.00	35.60	4.5	8.65	2.85	107.45
RW-13	7.19	329	218	8.1	210	12	79.3	13.33	11.36	0.51	42.00	33.20	3.6	9.27	3.11	97.75
RW-14	7.23	331	218	8.2	210	12	91.5	13.33	12.78	0.73	35.00	34.60	4.3	8.35	3.11	100.63
RW-15	7.14	334	219	8.2	210	9	97.6	10.00	11.36	0.93	56.00	32.20	4.5	8.65	3.37	98.96
RW-16	7.24	331	221	8.4	210	12	91.5	10.00	11.36	0.98	41.00	34.00	3.7	8.96	3.11	100.16
RW-17	7.23	330	224	8.4	210	12	109.8	6.66	12.07	0.82	19.00	36.60	4.6	8.65	3.11	110.36
RW-18	7.22	331	224	8.4	210	9	97.6	20.00	12.07	0.76	32.00	36.10	4.1	9.27	2.85	107.05
RW-19	7.23	331	224	8.4	210	12	85.4	6.66	10.65	0.58	34.00	31.60	3.4	8.65	3.11	92.93
RW-20	7.21	331	224	8.4	210	12	97.6	6.66	12.07	0.56	37.00	33.50	3.7	8.65	3.37	98.91
RW-21	7.18	328	225	8.5	210	9	97.6	13.33	11.36	0.78	67.00	33.80	4.2	8.65	2.59	101.72
RW-22	7.22	329	226	7.9	210	12	97.6	6.66	11.36	2.14	23.00	34.40	3.8	8.65	2.85	101.57
RW-1GW	7.27	516	214	8.1	350	18	134.2	6.66	9.94	14.55	5.00	42.60	6.60	21.02	1.55	133.56
RW-GW1	8.07	330	238	8.1	240	9	109.8	37.5	12.07	7.99	2.20	37.00	6.00	5.78	3.10	117.12
RW-GW2	7.42	480	238	8.5	320	12	134.2	18.75	12.78	9.33	17.60	47.00	10.00	9.27	2.36	158.57
RW-GW3	7.33	780	232	7.8	520	6	103.7	34.37	30.53	10.30	13.20	69.00	18.00	15.50	4.20	250.57
RW-GW4	7.35	490	233	7.6	320	6	109.8	12.50	11.36	12.30	4.40	43.00	7.00	23.00	2.33	137.84

H_2CO_3 and higher CO_2 may cause greater dissolution of minerals consequently releasing more of elemental ions in the water. The higher concentration of Na^+ and low Cl^- in the groundwater do not indicate presence of halite and/or evaporite in the sub-surface lithology. On the other hand, the Na- bearing alumino-silicate, such as plagioclase feldspar is usually present in sedimentary silicate rocks (shales, sandstones) which may have leached Na^+ and SiO_2 in excess under chemical influence of excess H_2CO_3.

The Rewalsar water is undersaturated with respect to calcite and dolomite and plots in undersaturated field of saturation indices diagram (Fig.5.20). The high partial pressure of carbondioxide $PCO_2^{10-2.09}$ compared to ambient $PCO_2^{10-3.5}$ of atmosphere, low pH and alkalinity are indication of CO_2 dissolution resulting from respiration and organic decay in the basin, holding $CaCO_3$ in solution and thus restraining calcite precipitation.

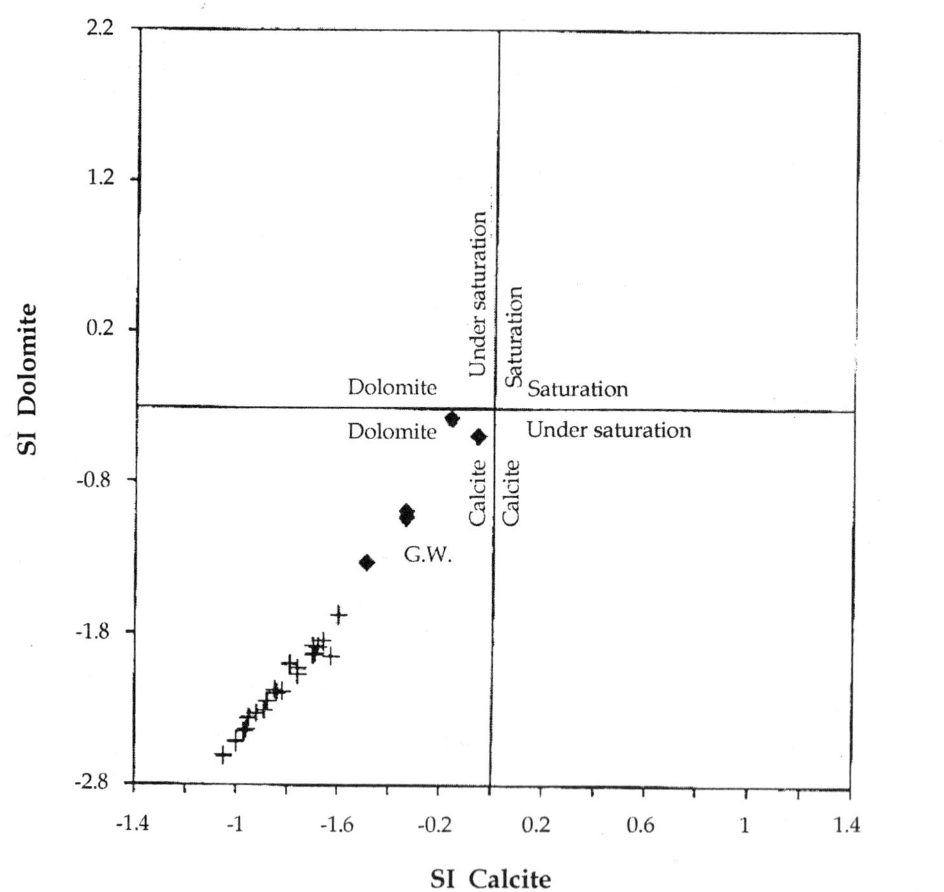

Fig. 5.20. Saturation Indices plot of calcite (SIc) vs dolomite (SId), the surface and ground water fall in undersaturated domain.

The Ca^{2+} + Mg^{2+} of 81.57% (on equivalent basis) of total cations, HCO_3^- + CO_3^- of 75.92% of total anions, represents higher dissolution and mobility of Ca^{2+} which alone constitutes 68.12% in water relating to those which are sourced from silicate rocks (Table 5.11). Although sedimentary silicates are the predominant lithology but the only possible source for the bulk of the calcium is from a phase calcite which releases only Ca^{2+} and HCO_3^- in solution. Very meager amount of calcite may be present in sandstone, shale matrix but it may be a dominant source of solutes during weathering (Drever, 1988). Minerals do not contribute to the solutes in runoff in proportion to their abundance in bedrock. The low ratio of Na^+ to Ca^{2+} does not support weathering of plagioclase feldspar to be the source of calcium in the water. Drever (1988) has pointed out that the weathering of plagioclase releases Ca^{2+} and Na^+ to solution in approximately the same ratio as they occur in solid.

Table 5.11: Percentage of equivalent per million of the constituents in Rewalsar Lake

Location	Cations			Anions		
No.	Ca	Mg	(Na+K)	SO_4	Cl	(CO_3+HCO_3)
RW-1	67.97	13.28	18.75	16.80	11.20	72.00
RW-2	67.10	13.42	19.48	18.10	12.93	68.97
RW-3	67.07	14.06	18.87	8.57	13.88	77.55
RW-4	68.29	13.01	18.70	11.57	14.05	74.38
RW-5	68.34	13.51	18.15	10.61	13.63	75.76
RW-6	68.24	14.51	17.25	13.21	16.04	70.75
RW-7	67.83	13.48	18.69	6.19	15.16	79.65
RW-8	67.60	13.60	18.80	13.94	14.34	71.71
RW-9	67.83	14.34	17.83	10.69	12.98	76.33
RW-10	68.29	13.01	18.70	11.57	14.05	74.38
RW-12	68.73	14.28	16.99	11.20	12.80	76.00
RW-13	68.73	12.40	19.01	12.17	13.91	73.91
RW-14	67.58	13.67	18.75	11.02	14.17	74.80
RW-15	66.53	15.29	18.18	8.64	13.17	78.18
RW-16	69.11	12.19	18.70	8.64	13.17	78.19
RW-17	68.28	14.18	17.54	5.22	12.69	82.09
RW-18	69.62	13.08	17.31	15.79	12.78	71.43
RW-19	67.52	11.97	20.51	6.25	13.39	80.36
RW-20	68.57	12.65	18.78	5.64	13.71	80.65
RW-21	68.15	14.11	17.74	11.20	12.80	76.00
RW-22	69.35	12.50	18.15	5.69	13.01	81.30
RW-11GW	58.68	15.15	26.17	4.35	8.69	86.96
RW-GW1	69.55	18.80	11.65	28.68	12.50	58.82
RW-GW2	65.28	23.05	11.67	11.64	10.75	77.61
RW-GW3	60.85	26.45	12.70	14.57	18.62	66.80
RW-GW4	50.52	15.27	24.21	11.47	9.41	79.12

The (Ca +Mg): (Na +K) mol ratio 2.2 suggests that the major ions contribution by silicate weathering is less significant. This is supported from data on Ca^{2+}/Na^{+} mole ratio ~ 2.23, Mg^{2+}/Na^{+} ~ 0.44 and HCO_3^- 3.87 which are much lower than the water that drains carbonate lithology but relatively higher than those draining silicate rocks.

This shows that the elemental ions are derived from both silicate and carbonate sources. In case of carbonate weathering the molar ratio of 0.50 as the stoichiometric relationships of both Ca^{2+}/HCO_3^- and $Ca^{2+} + Mg^{2+}/ HCO_3^-$ has been reported by Zhang and others (1995). This ratio in Rewalsar Lake is 0.58 and 0.69, respectively, indicating involvement of silicate weathering in the basin.

The plagioclase feldspar albite (Na-plagioclases) is common in sedimentary rocks and easily undergoes weathering. With carbonic acid: $2H_2CO_3 + 9H_2O + 2NaAlSi_3O_8 = Al_2SiO_2O_5(OH)_4 + 2Na^+ + 2HCO_3^- + 4H_4SiO_4$ (Garrels,1967). This is illustrated by stability diagram using log $[Na^+/H^+]$ and log $[H_4SiO_4]$, the Rewalsar water samples plot on kaolinite-gibbsite boundary as well as in gibbsite field and also in log K vs. H_4SiO_4 diagrams (Figs. 5.21a & 5.21b).

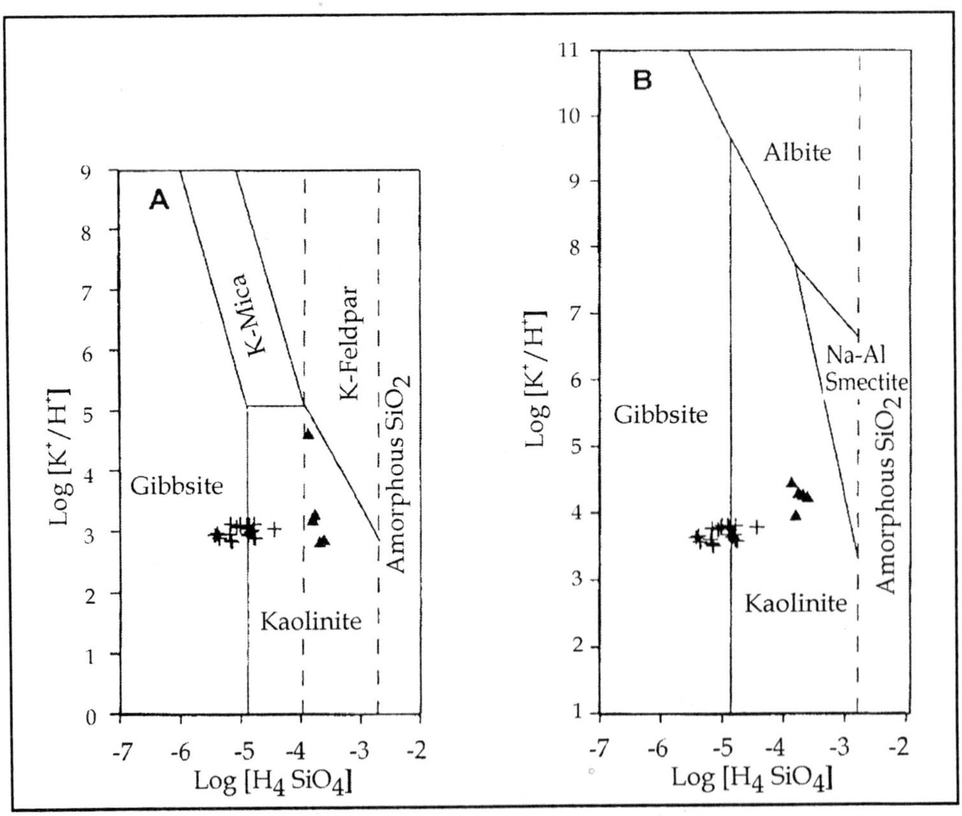

Fig. 5.21 (a & b) In stability diagrams (a) & (b) surface water plots in gibbsite field whereas ground water fall in kaolinite field.

The removal of both cations and silica due to rapid flushing, allows shorter time of contact with the primary minerals, resulting in the lower dissolved concentration in the existing waters. Rapid flushing due to rainfall, good drainage due to high relief have removed cations and silica. And additionally, the latter is further depleted due to silica depletion, and uptake by diatoms and micro-organisms. On account of these, the diagram demarcates the regions of stability of albite in solution led water samples of Rewalsar to fall in gibbsite field whereas the groundwater samples which are rich in silica and Na^+ plot in kaolinite field. In a log Mg^{2+} $+Ca^{2+}$ vs H_4SiO_4 plot, the water samples lie in gibbsite field and groundwater in kaolinite field (Fig.5.22a and b).

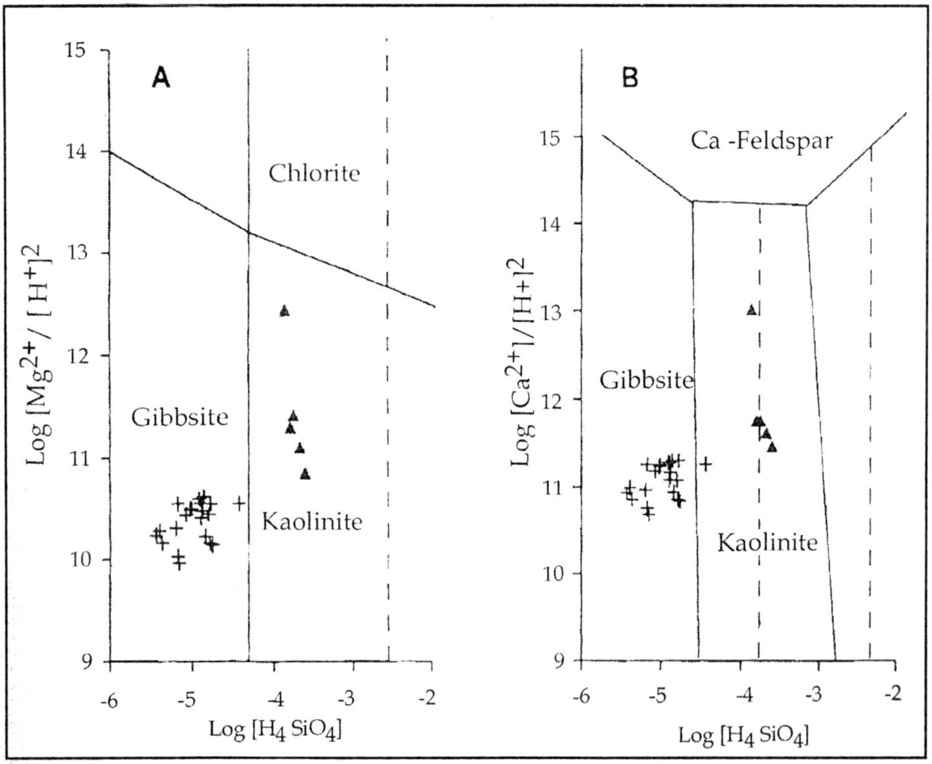

Fig. 5.22 (a) and (b) In a plot of Mg^{2+} and Ca^{2+} vs H_4SiO_4 log values, surface water lies in gibbsite field while groundwater in kaolinite field.

This is consistent with the general concept and illustrates that deficiency of silica and cation leads to the formation of gibbsite and with their increasing concentration kaolinite and smectite are in turn formed (see Drever, 1988). Therefore, Rewalsar Lake is a good example of such a process to have taken place.

There is a considerable variation in SO_4^{2-} concentration in surface water (6.66 to 20.00 mg/l) and groundwater (6.62 to 37.5 mg/l) at different locations. The main source of SO_4^{2-} is sedimentary lithology and anthropogenic input.

As the sulphate-bearing minerals are not reported from rock formation the variation may be related to anthropogenic sources. The groundwater samples collected from more polluted sites are generally in excess of SO_4^{2-}. On the other hand, Cl^- content is 13.52 mg/l on an average in lake water without much variation, and in groundwater its abundance is 15.33 mg/l with a higher Na^+ content 14.90 mg/l as compared to 8.72 mg/l in surface water indicating greater dissolution of plagioclase from sub-surface lithology. There is a negative correlation of Cl^- with Na^+ in surface water samples indicating source other than halite or sea salt aerosole. On the other hand, Na^+ - bearing alumino-silicate, such as plagioclase may have leached Na^+ and SiO_2 in excess under the chemical influence of excess H_2CO_3 in ground water.

Kunti Bhayog Lake

The lake is situated at the hill top of Rewalsar town, Mandi district at an altitude of 1750 m above msl between 31°37' N latitude and 76°49' 6" E longitude. It is a triangular, fresh water lake 8 Km by road from Rewalsar. The habitation is extremely sparce and the lake maintains its pristine beauty (Fig.5.23). The lake basin is formed in sedimentary formation comprising siltstone (grayish-greenish in colour), hard compact and interbedded with sandstone of middle Siwalik Formation. It is weathered and show development of mud cracks, rain prints and typical spheroidal weathering of sandstone along the banks of the lake (Fig.5.24). The sedimentary structures speak of shallow water deposition of the rocks. The sandstone is greywacke.

Fig. 5.23. A panoramic view of Kuntibhayog Lake showing catchment lithology, sparce vegetation and habitation.

Fig. 5.24. Middle Siwalik showing spheroidal weathering of sandstone.

The crushing of rocks and abrupt changes in dip, steep and gentle may indicate a fault in the area. This lake, like that of to Rewalsar may be of tectonic origin.

The lake is moderately alkaline with pH ranging between 7.20 to 7.36 and HCO_3^- 39.69 mg/l and Ca and Mg 13.5 and 1.75 on average, respectively (Table 5.12). This has resulted in low TDS 70 mg/l and low E.C. 105 µs/cm. The low Ca and Mg has yielded low alkalinity 0.8 meq/l. Similarly, low dissolved oxygen (D.O) 5.2 mg/l (on average) indicates decomposition of organic matter and respiration in the basin. The water is clear and greenish due to blue-green algae. However, the photosynthetic activity is retarded due to respiration lowering pH and dissolved oxygen in water of the lake. The partial pressure of carbondioxide log $PCO_2^{-3.84}$ is lower than atmospheric $PCO_2^{-3.5}$ which supports low Ca and Mg dissolution in water and consequently low CO_2 in water.

The silica content is quite low 0.305 mg/l which is also supported by the poor presence of Bacillariophyceae (Singh et al., 1986). This, however, is not in conformity with the catchment lithology of the Kunti Bhayog Lake comprising predominantly of sedimentary silicates viz. sandstone, siltstone, rocks. Low dissolution of sedimentary rocks comprising detrital minerals quartz, mica and clays that are resistant to weathering releases less silica. Also because of low to moderate alkalinity (0.775 meq/l) indicating low CO_2 and consequently low carbonic acid formation rendering water a weaker

Table 5.12: Chemical parameters of Kunti Bhayog Lake, Himachal Pradesh (conc. in mg/l)

Parameters	pH	E.C (μs/cm) at 25°c	DO	ORP (mv)	CO_3	HCO_3	Cl^-	SO_4^{2-}	Ca	Mg	Na	K	PO_4^{2-} (μg/l)	SiO_2	TDS	TH
Ku 1	7.34	110	5.4	210	3	24.4	15.62	18.75	11.00	2.00	2.74	1.99	15.40	0.260	70	35.75
Ku 2	7.36	100	5.3	211	6	42.7	14.20	12.50	14.00	2.00	2.37	1.99	11.00	0.312	70	43.20
Ku 3	7.20	110	4.7	215	3	30.5	13.49	12.50	11.00	1.00	2.67	1.66	11.00	0.273	70	31.59
Ku 4	7.26	100	5.6	214	3	61.0	12.07	12.50	18.00	2.00	2.97	1.99	4.40	0.377	70	53.20

Percentage of equivalent per million of constituents in Kunti Bhayog Lake

	Ca^{2+}	Mg^{2+}	Na+K	SO_4^{2-}	Cl^-	$CO_3^-+HCO_3^-$
Ku 1	68.75	10.00	21.25	29.32	33.08	37.59
Ku 2	67.96	16.50	16.50	16.67	25.64	87.69
Ku 3	62.50	19.32	23.86	20.97	30.64	48.37
Ku 4	72.58	13.71	13.71	20.00	15.29	64.70

solvent. Although it is not always, but total dissolved silica may be high in alkaline waters. The very low concentration of Na^+ & K^+ 2.69 and 1.90 mg/l average, respectively also indicates poor dissolution of Na- plagioclase and K-feldspar. The former is commonly present in sandstone and sedimentary silicate rocks. The concentration of Na and K is almost 4 and 2 times respectively, which is less than Rewalsar Lake. Similarly, Ca and Mg are 3 times lower than Rewalsar and bicarbonate is 4 times less.

Lake Kunti Bhayog, though located in the same geological and climatic region as Rewalsar, shows variable physico-chemical characteristics. This is because of predominance of a particular rock type and absence of other in the watershed areas of the two. The thin discontinuous layer of limestone in Rewalsar lithology is absent in this lake and a thicker sequence of sandstone-silt-clay rocks predominate in Kunti Bhayog Lake catchment with increasing height. Also anthropogenic interference in Rewalsar is relatively much higher than in this lake. The SiO_2 content is 2 times less but Cl- and SO^{2-}_4 concentrations are almost similar and PO_4^{2-} content is little higher in Rewalsar. Such a variation in chemistry of the two lakes seems to be controlled by human activity, biological changes and rock types. These lakes are example of rock types controlled chemistry of water. Among the siliciclastic (detrital) rocks, shales are more complex. The most abundant minerals in shales, illite and quartz, are relatively unreactive in the weathering environment and contribute relatively little to runoff chemistry (Drever, 1988). The water draining shale-siltstone in Kunti Bhayog Lake has sulphate and chloride in equal percentage and contribute 50% of the total anion budget. The Cl^- shows negative correlation with Na^+, and SO_4^{2-} negative correlation with Ca^{2+}, suggesting that they are not derived from halite and/or gypsum, respectively. The shales may be containing pyrite, which may contribute sulphate to solution after oxidation. Another source of SO_4^{2-} is human activity, which is not significant in this lake. The concentration of dissolved inorganic phosphate (PO_4^{2-} = 10.45 mg/l) is higher as compared to available phosphorus responsible for rich phytoplankton population (Singh et al., 1986). Pollution indicator species Microcystis commonly occur in this lake, which is also reported as pollution indicator in various lakes by Hutchinson (1967), Prescott (1968) and Nygarrd (1949). However, species Microcystis is also reported from lakes, which are not an organically rich lake.

Khajjiar Lake

The Lake Khajjiar at an elevation of 1927 m above msl is located in lust green meadow, in the midst of dense deodar (cedrus deodara) forest and is about 13 kms by road from Dalhousie – a hill station of Himachal Pradesh. The lake is oval shaped, surrounding wide marshy zone which is followed by a meadow, measuring about 2.5 Km in its periphery and is 6 m deep (Fig.5.25).

Its typical shape and location in the meadow led many scientists to speculate that it has formed due to huge meteorite fall. The lake is in district Chamba within the limits of 32° 33' N latitude and 76°3' 5" E longitude. Previous studies suggested that the lake covered a much larger area than it is today in about 4000 year BP (Sharma 1970; Sharma and Singh, 1972). A special feature of the lake is that the detritus at the bottom of the lake, has been estimated to be upto 2 m thick, consisting mostly of leaf fragments of phragmites and Acorus. During the summer, large amounts of detritus rise to the surface to form loose aggregates which gradually becomes consolidated and are colonized by grasses and sedges. A few years ago (2001) a part of that has been removed by the Himachal Government to expose more water area of the lake to attract tourists.

Fig. 5.25. Oval-shaped Khajjiar Lake located in a lust green meadow surrounded by deodar forest-known as mini Switzerland.

The lithology around the lake is coarse grained Dhauladhar granite. Near the contact with the country rock it has generally developed a gneissic structure. The main granitoid rocks have the composition of granodiorite (quartz,K-feldspar,Na-plagioclase and mica) locally changing to quartz diorite.

The water chemistry of Khajjiar Lake is rather different from the other lakes of lesser Himalaya. The pH of water is low, 6.05 (on average) and 6.58 in the open nullah flowing across the meadow and 7.01 in groundwater indicating that the water is neutral. This is due to the fact that Ca^{2+} content is very low 12.20 mg/l and Mg^{2+} 0.38 mg/l. Bicarbonate (HCO_3^-) is 34.4 mg/l and CO_3^{2-} in traces because at neutral pH carbonate exists in the form of HCO_3^-, which is the dominant anion indicating that the water is devoid of, carbonate minerals (Fig.5.26; Table 5.13). The diagram demonstrates the role of pH in controlling HCO_3^- and CO_3^{2-} in water. A low Ca + Mg: HCO_3 equivalent ratio 1.23, (n=9) and low CO_3^{2-} + $HCO3^-$ alkalinity 0.51 mg/l av. indicating low carbonate dissolution in water, contrary to this in groundwater Ca + Mg : HCO_3 ratio is 0.86 on equivalent basis and alkalinity is 1.20 meq/l, this is because the HCO_3^- ion concentration is higher than Ca + Mg combined together on equivalent basis indicating contribution of silicate source rocks. The Ca + Mg/ Tz^+ ratio 0.92 in lake and 0.8 in groundwater indicate silicate source contributing cations and anions in lake and groundwater. However, low Na + K/Tz^+ ratio 0.075 does not indicate significant contribution from silicate source rock. The HCO_3^- show strong positive correlation with Na (0.82), moderate with Mg and low negative correlation (-0.22) with Ca indicating silicate source. Chloride shows strong positive correlation with Na and Mg and very weak with K and Ca. This indicates that Cl^- is derived from precipitation. The atmospheric CO_2 in water is converted to organic matter by green plants deriving energy from photosynthesis. However, decay of phragmites, biomass and phytoplankton caused respiration in the basin. This has resulted in high log $PCO_2^{-1.46}$ compared to ambient $PCO_2^{10-3.5}$ atm. Waters with PCO_2 values in this range in equilibrium with calcite would have pH 7.3 and 8.4 (Drever, 1988). The lower pH in Khajjiar Lake water indicates that it is undersaturated with respect to calcite and dolomite, hence plot in undersaturated domain of saturation indices diagram (Fig.5.27). In such waters, the contribution of solutes from the atmosphere may be as large as the contribution from rock weathering and thus the weathering of minor phase such as calcite may be the principle source of solutes (Drever, 1988). This is indicated from the relatively higher content of Ca^{2+} among cations and HCO_3^- among anions in the lake water. Nevertheless, the silicate weathering is well exemplified in the sediment chemistry of this lake. The bed sediments are distinctly enriched in Si, Al, Fe, K, Mg and Ti, while Ca is highly depleted. This may reflect weathering of granite or grano-diorite in the catchment area of the lake. This data when compared to the sediment chemistry data of other lakes of Himalaya which have carbonate dominated rocks in their catchment lithology have these elements much lower, and Ca, Sr etc. higher than, in sediments of Khajjiar Lake. Low pH, E.C and TDS in waters indicate weaker dissolution capacity of solutes. Therefore, such waters are weak in H_2CO_3 and thus minerals do not contribute to the solutes in proportion to their abundance in the bed rock (Drever, 1988).

The water is highly depleted in PO_4^{3-} as the plants require nitrogen, phosphorous compounds and a wider range of trace elements. Plants like microscopic algae in aquatic systems will grow and multiply until all available

Table 5.13: Chemical Characteristics of the Khajjiar Lake waters (Conc. in mg/l)

Parameters	pH	E.C. (μs/cm) at 25°c	ORP (mv)	DO	CO_3^{2-}	HCO_3^-	Cl^-	SO_4^{2-}	PO_4^{3-}	SiO_2	Ca^{2+}	Mg^{2+}	Na^+	K^+	TDS	TH
Kj 1	6.20	146	73	7.7	nil	36	5.68	1.2	0.006	10.75	7.19	0.316	0.96	1.18	102	19.26
Kj 2	6.07	27	180	8.5	nil	46	5.68	1.63	0.075	3.83	19.20	0.494	0.93	0.59	19	49.99
Kj 3	6.15	26	174	9.1	nil	48	5.68	1.39	0.069	3.56	12.70	0.381	0.93	0.59	18	33.29
Kj 4	6.02	22	193	9.3	nil	36	5.68	0.81	0.031	3.56	11.71	0.369	0.69	0.59	15	30.77
Kj 5	6.02	27	178	8.8	nil	36	4.26	0.6	0.019	3.27	11.44	0.365	093	0.59	19	30.08
Kj 6	6.11	21	213	9.6	nil	24	5.68	1.63	0.012	3.50	11.63	0.372	0.93	0.59	15	30.58
Kj 7	6.08	25	230	9.1	nil	24	7.10	0.91	0.037	3.36	12.05	0.398	0.93	1.18	17	31.74
Kj 8	6.02	27	224	10.1	nil	24	5.68	2.73	0.069	3.58	12.28	0.401	0.93	0.59	19	32.33
Kj 9	6.06	26	231	9.3	nil	36	5.68	0.91	0.006	3.53	11.67	0.380	0.93	0.59	18	30.72
Kj 10(nala)	6.81	61	230	7.8	nil	36	7.10	1.68	0.101	6.64	20.59	0.801	0.93	0.59	43	54.73
Kj 11(nala)	6.36	54	221	8.0	nil	24	7.10	1.25	0.120	4.90	20.75	0.831	0.93	0.59	38	52.25
Kj12 Handpump	7.01	204	265	7.6	nil	72	9.94	3.02	0.224	11.60	18.63	1.058	5.33	0.59	143	50.89

phosphate or nitrate are used up at which point photosynthesis ceases. In unpolluted fresh waters, phosphate is normally the limiting nutrient. This may explain very low PO_4^{3-} content in water of Khajjiar Lake.

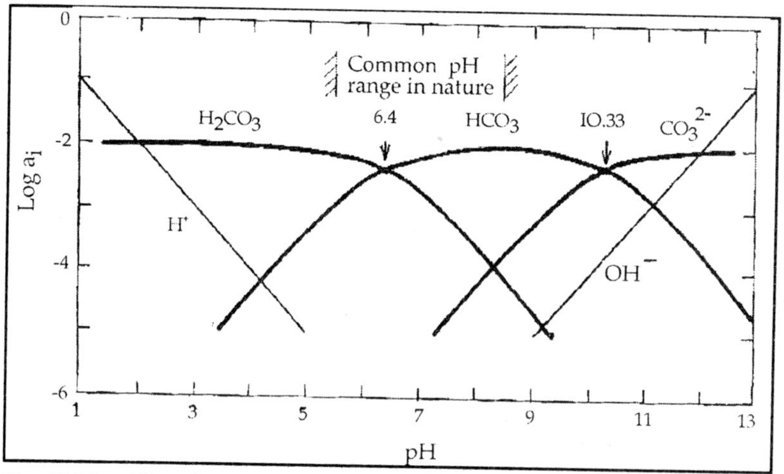

Fig. 5.26. pH vs Log ái plot showing pH control of H_2CO_3, HCO_3, CO_3^{2-} (after Drever, 1988).

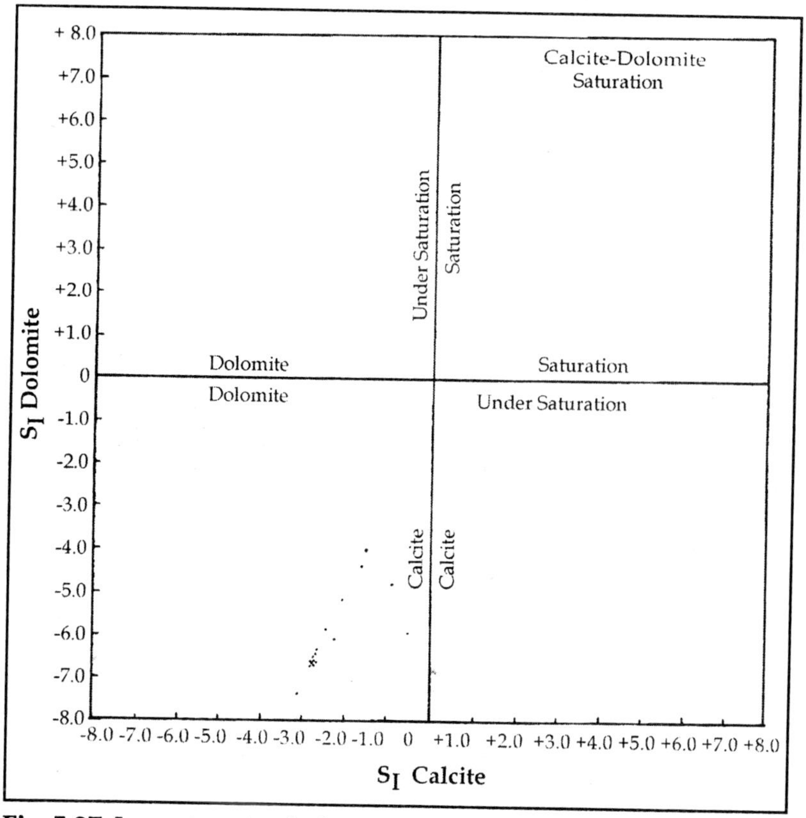

Fig. 5.27. In a saturation Indices plot the water chemistry of Khajjiar lies in undersaturated domain.

Lakes of Kumaun Himalaya

Nainital Lake

A beautiful crescent shaped lake with the hill slopes studded with uncountable houses/buildings and often low floating clouds over the hills, sometimes descending and touching the lake, is a magnificent beauty of this lake which attracts large number of tourists from far and wide (Fig.5.28 a, b).

Fig. 5.28 (a) A panoramic view of crescent shaped Nainital Lake which is densely populated in the valley and hill slopes

Fig. 5.28 (b) A close-up view of Naini Lake showing very dense settlements on the hill slopes

The lake is about 1.5 km long and 0.36 km wide located in the valley at an altitude of 1938 m from msl and is within the geographic limits of 29^0 22' 62" N and 79^0 27' 37" E longitude. The bathymetry has shown a median ridge about 100 m wide, running transversely dividing the lake into two sub-basins- Talli-Tal in the NW and Mallital in SW. The maximum depth of the basins are 27m and 25 m, respectively. The depth varies from 11 to 25 m in Talli Tal sub-basin and 7 to 27 m in Malli Tal sub-basin. The maximum thickness of sediment recorded with the help of ORE sub-bottom profiler (3.5 KHZ frequency) extends upto 15 m in both the sub-basins (Rawat 1987; Jauhari and Hashimi, 1994). It is a fresh water lake formed due to tectonic activity during Holocene time. The Naini Lake originated as a result of differential vertical-rather-rotational movements along the Nainital Fault leading to blocking of a perennial stream (Balia River) in its upper reaches resulting in the formation of Naini Lake (Valdiya 1988). It is related to neotectonic movements on the Main Boundary Thrust and with the genetically related faults (Fig.2.2). The main feeder stream Bara Nala emanating from the debris fans at the foot of the Naina Peak and derives larger part of its discharge from Parda spring which is located on the Nainital fault of great extent and depth. A large number of gullies descending almost transversely down the steep slopes carry discharges from the springs. The inlet nullahs passing through the market and habitation carries untreated polluted sludge which drains off into the lake. The annual rainfall is 249 cm. The valley has about 40,000 inhabitants and is witnessing a rapid change in its ecological system due to anthropogenic activities. This has entailed deforestation resulting in silt flow and pollution in the lake. The lake has 22 inlets nullahs (streamlets), nine of which are permanent and act as the major conduits of polluted sludge and silt draining into the lake. The rate of sedimentation is measured as 11.4 mm/yr by Pb^{210} method (Das *et al.*, 1994) which is very high. Nainital experiences prolonged winter extending from October to March and sometimes temperature goes down below freezing. The lake is warm monomictic type with a winter homothermal profile of $7\text{-}8^0$ C and a weak stratification in summer with 20^0 C in surface layers and 12^0 C in deep water. The catchment of the lake is comprised of Krol Formation (Precambrian) consisting predominantly of dolomitic limestone, calcareous slates, ferruginous shales, greywackes, siltstone, gypsum etc. (Fig. 5.29).

The Nainital Lake water is mildly alkaline and pH ranges from 7.25 to 8.97 with an average of 8.67. The E.C of surface water ranges from 670 to 945 μs/cm with an average of 706 μs/cm, correspondingly, the TDS ranges from 430 mg/l to 610 mg/l with an average of 441mg/l. This has lead to higher ionic strength (13.1×10^{-3} m mols/l) of lake water. The EC and TDS of Nainital Lake water is controlled by high dissolution of limestone and dolomite which are the predominant lithology of the lake catchment area (Table 5.14) (Singh and Das, 1995; Das, 2005). The highest concentration is recorded at the sampling point where Chuna Dhara spring drains the lake. Magnesium is the most dominant cation with the average of 68.5% which is 2.5 to 3 times more than Ca^{2+}. Among anions, HCO_3^- is the most dominant, averaging about 73.13% followed by SO_4^{2-} averaging 22.13% and chloride 4.73% (Table 5.15).

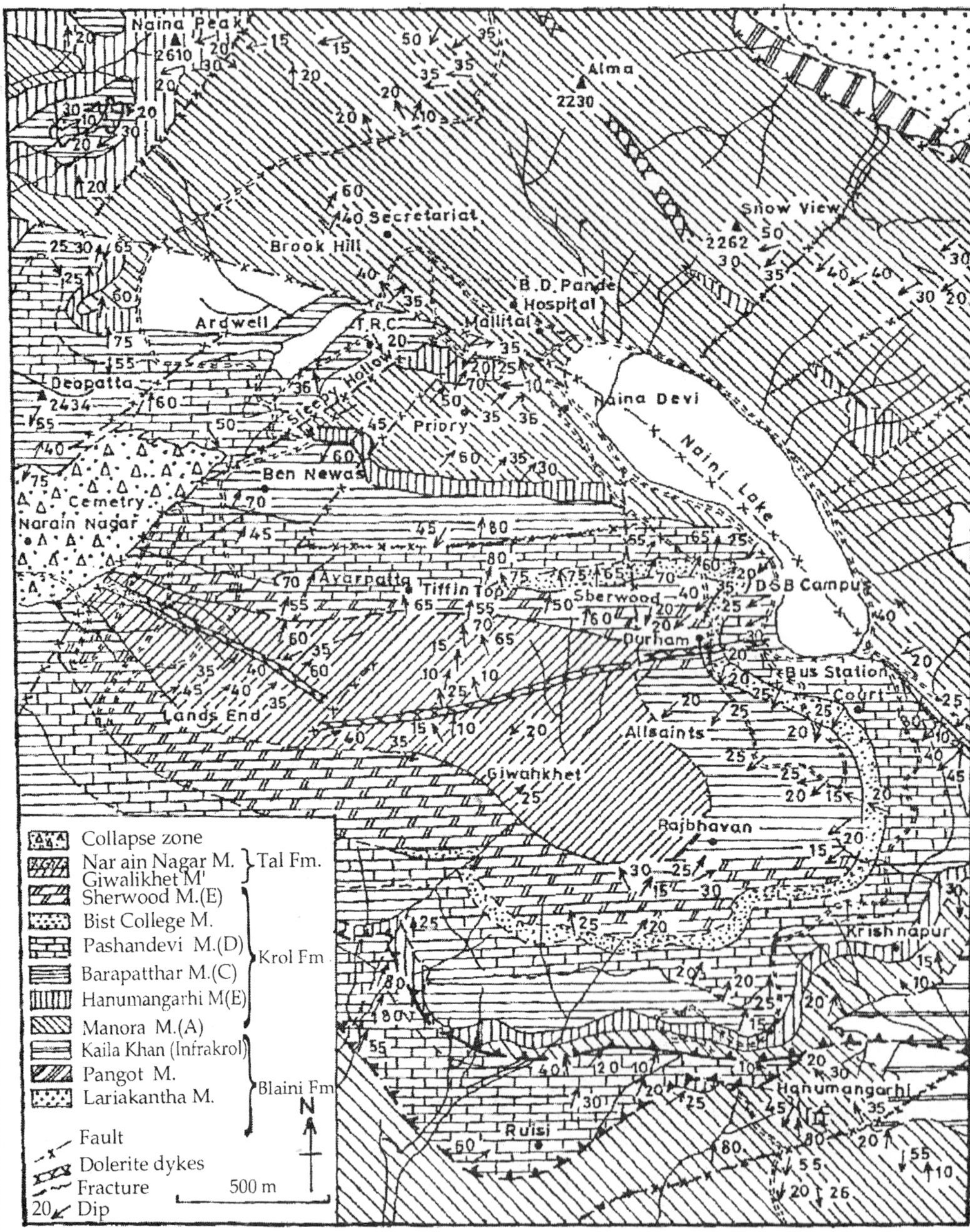

Fig. 5.29. Geological and tectonic map of Nainital showing lithology and structure around Nainital (after Valdiya, 1982).

Table 5.14: Chemical characteristics of Nainital Lake, Kumaun Himalaya

Sample No.	pH	E.C. (µs/cm)	DO	CO_3^{2-}	HCO_3^-	Cl^-	SO_4^{2-}	PO_4^{3-}	SiO_2	Na^+	K^+	Ca^{2+}	Mg^{2+}	T.H	TDS	TSM
									mg/l							
1.	8.80	723	11.3	36	390.4	16.23	81	0.225	3.68	14.52	3.91	41.68	57.95	345.99	440	1.2
2.	8.02	945	10.6	24	366.0	28.40	103	0.335	7.12	28.08	7.88	47.29	76.45	437.19	610	0.9
3.	8.44	699	9.4	12	341.6	14.91	86	0.122	3.36	12.67	3.45	29.65	62.09	333.13	440	1.4
4.	8.39	703	8.3	12	402.6	14.20	67	0.113	3.36	12.36	3.45	20.04	60.38	301.95	440	0.9
5.	8.54	707	9.0	12	341.6	14.20	93	0.095	3.26	12.36	3.45	30.86	61.60	334.13	440	1.1
6.	8.44	705	7.9	24	341.6	15.62	113	0.085	3.18	12.67	3.45	31.26	61.84	336.14	450	1.5
7.	8.43	705	9.0	12	341.6	13.49	86	0.103	2.99	12.36	3.68	40.08	55.51	331.83	450	1.6
8.	8.44	707	8.5	66	353.8	18.46	103	0.095	3.16	12.36	3.45	27.25	63.06	331.83	460	0.9
9.	8.43	708	9.1	12	366.0	14.91	70	0.125	3.66	12.67	3.45	33.66	61.87	342.29	460	1.1
10.	8.43	708	8.2	36	329.4	14.20	93	0.120	3.20	12.36	3.45	41.28	50.67	314.66	450	1.3
11.	8.90	714	8.5	12	366.0	14.20	96	0.085	3.15	12.67	3.68	36.87	56.97	329.83	460	1.2
12.	8.92	690	8.9	60	347.7	15.62	83	0.162	2.97	12.98	3.45	34.46	63.78	343.14	440	4.0
13.	8.85	682	10.1	12	341.6	14.20	89	0.103	3.50	12.36	3.45	30.86	61.85	335.13	450	1.2
14.	8.76	678	8.4	12	414.8	14.20	105	0.083	3.50	12.67	3.45	41.68	54.54	331.78	430	1.1
15.	7.25	691	8.7	24	317.2	15.62	126	0.065	3.44	12.67	3.45	57.71	43.83	327.18	430	1.1
16.	8.82	695	8.9	78	353.8	14.20	89	0.113	3.48	12.67	3.45	31.26	60.87	336.14	440	1.4
17.	8.86	696	8.6	00	366.0	14.20	113	0.092	3.00	12.36	3.45	32.86	42.85	260.91	440	1.0
18.	8.75	693	8.3	60	366.0	15.62	103	0.040	2.94	12.67	3.45	34.06	50.40	295.45	440	2.3
19.	8.82	698	9.1	12	372.1	14.20	103	0.113	3.22	12.67	3.68	40.48	56.29	336.04	440	0.4
20.	8.85	707	8.8	12	341.6	14.20	83	0.103	3.27	12.67	3.45	30.86	51.11	337.64	440	1.1
21.	8.86	685	9.0	48	329.4	14.20	100	0.150	3.20	12.36	3.45	13.83	70.37	328.07	430	2.1
22.	8.86	684	9.6	60	305.5	14.91	100	0.138	3.46	12.67	3.68	20.44	63.31	315.17	440	0.8
23.	8.86	699	9.7	36	317.0	14.91	100	0.168	3.24	12.36	3.68	27.25	65.74	342.34	450	1.4
24.	8.93	687	9.5	48	305.0	16.33	115	0.136	3.17	12.67	3.68	12.42	70.37	327.57	430	1.3
25.	8.95	704	10.2	24	341.6	15.62	115	0.125	3.44	12.36	3.45	37.67	60.38	346.05	440	2.2
26.	8.95	702	9.0	12	353.8	14.20	126	0.113	3.53	12.67	3.45	37.67	59.17	340.99	440	1.3
27.	8.95	694	8.9	36	329.4	14.91	79	0.103	3.17	12.36	3.45	30.86	60.38	335.13	430	1.8
28.	8.97	700	8.7	12	341.6	14.20	100	0.136	3.64	12.67	3.68	29.25	56.70	313.71	440	2.8
29.	8.94	670	8.8	72	384.3	14.20	115	0.148	3.60	12.36	3.45	25.65	59.37	311.76	430	1.8

Table 5.15: Percentage of Equivalent per million of the constituents in Nainital Lake

Sample No.	..............Cations...............			..............Anions...........		
	Ca	Mg	(Na+K)	(CO_3+HCO_3)	SO_4	Cl
1.	27.26	63.21	9.53	77.93	17.30	4.77
2.	23.31	62.70	13.99	69.72	22.00	8.28
3.	20.37	70.50	9.13	73.04	21.81	5.15
4.	15.04	75.50	9.46	79.56	15.86	4.58
5.	21.13	70.30	8.57	71.94	23.23	4.83
6.	21.25	70.00	8.75	69.59	25.59	4.82
7.	27.59	63.71	8.70	73.40	21.91	4.69
8.	19.70	76.00	4.30	74.99	20.11	4.90
9.	22.50	68.90	8.60	77.28	17.61	5.11
10.	29.86	61.00	9.14	73.82	21.67	4.51
11.	25.47	65.60	8.93	72.71	22.72	4.57
12.	22.94	68.30	8.76	77.99	17.51	4.50
13.	21.24	70.90	7.86	72.67	22.45	4.88
14.	28.67	62.50	8.83	73.55	22.34	4.11
15.	40.20	50.80	9.00	66.17	28.95	4.88
16.	21.25	70.00	8.75	78.82	17.39	3.79
17.	28.14	61.15	10.71	68.52	26.88	4.60
18.	26.23	64.70	9.07	75.56	20.26	4.18
19.	27.49	63.70	8.81	71.83	23.71	4.46
20.	21.03	70.20	8.77	73.78	21.26	4.96
21.	9.62	81.60	8.78	73.78	21.96	4.26
22.	14.72	75.90	9.32	73.63	21.91	4.46
23.	18.23	73.30	8.43	71.85	23.38	4.77
24.	8.71	82.20	9.09	69.77	25.32	4.91
25.	24.97	66.70	8.33	69.27	25.93	4.80
26.	25.33	66.30	8.37	67.19	28.44	4.37
27.	21.07	70.30	8.63	76.14	18.98	4.88
28.	22.32	68.30	9.38	70.70	24.54	4.76
29.	18.70	72.17	9.13	75.65	20.83	3.52

Therefore the lake water is HCO_3>SO_4 >Cl type (Fig.5.30) and is similar to the trend reported from world water bodies. But cation concentration Mg > Ca >Na >K deviate from the general trend Ca >Mg>Na > K. This shows that the water is draining through dolomite and dolomitic limestone of the catchment area. The sulphate content in lake water is derived from the pockets of gypsum associated with limestone in the valley.

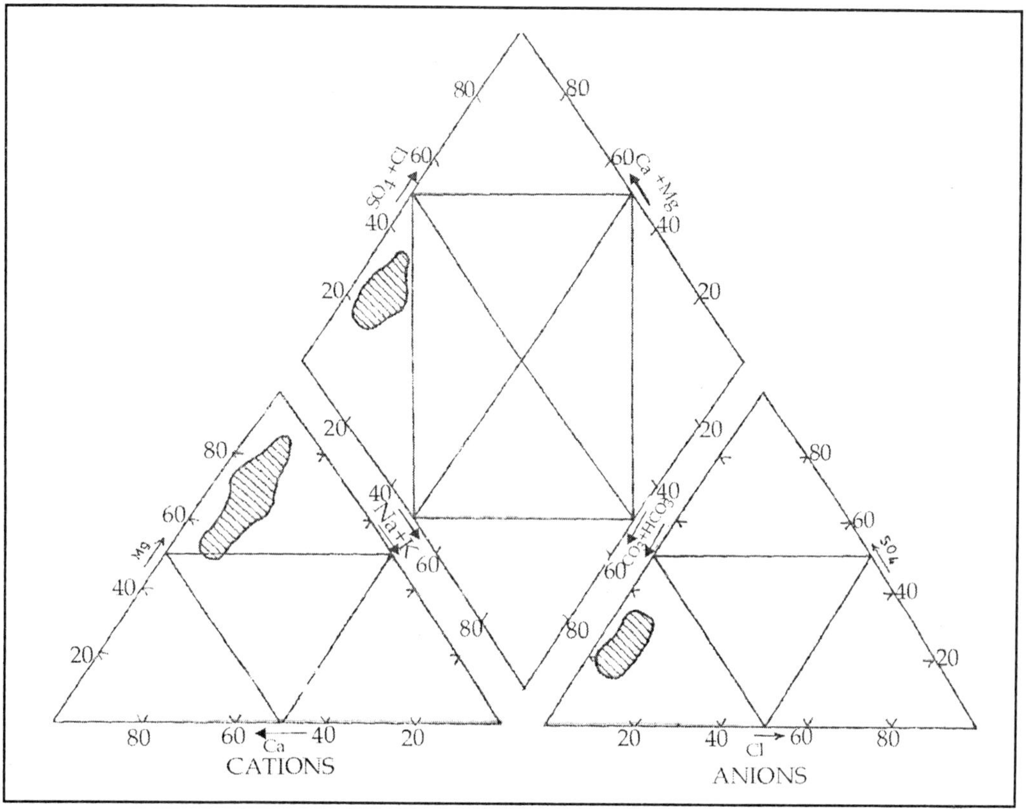

Fig. 5.30. In a trilinear diagram the Nainital Lake water represents Mg - HCO_3 - SO_4 type facies.

This is supported from the positive correlation of sulphate with calcium (r = 0.17) indicating a common source. The abundance of Cl^-, Na^+ and K^+ are low in the lake water. This is because the lake is far away from the sea. Hence, the probability of sea salt aerosol reaching as far as Nainital in the north is improbable, and the salt deposits are also not reported from this region. Hence the rock formation and input from anthropogenic activity are the probable sources. The weathering of lithology is a important mechanism for the release of ions from the constituent minerals to the water, therefore, the plots of TDS against (Na/Na + Ca) on Gibbs (1970) diagram (Fig. 5.31), the water chemistry of Nainital Lake falls in the rock dominant domain. The relative abundance of Na^+ over K^+ in this lake is due to the greater mobility of Na^+ during weathering. K^+ being less mobile and less soluble remains in the residue. Reverse weathering during which potassium is utilized in the formation of clay minerals like illite (Von Damm and Edmond, 1984) may also deplete K^+ in the water pool. The equation $(Na^+ + K^+)/T_Z^+$ is generally considered as an index to assess the contribution of cations derived from silicate weathering, where Na^{+*} is taken after correction for cyclic salts (Stallard and Edmond, 1983, 1987). The low ratio (0.09) in Nainital Lake indicates that there is no significant contribution of cations from silicate weathering in this lake. The high (Ca + Mg): (Na + K) equivalent ratio 10.14 and (Ca + Mg): HCO_3 equivalent ratio 1.16, show that the latter satisfy the stoichiometry requirement of carbonate weathering hence contributing predominantly the element ions to the water. Ca + Mg / T_Z^+ ratio 0.9 on average may support carbonate weathering as well in Nainital Lake (Sarin *et al.*, 1989; Stallard and Edmund 1983, 1987).

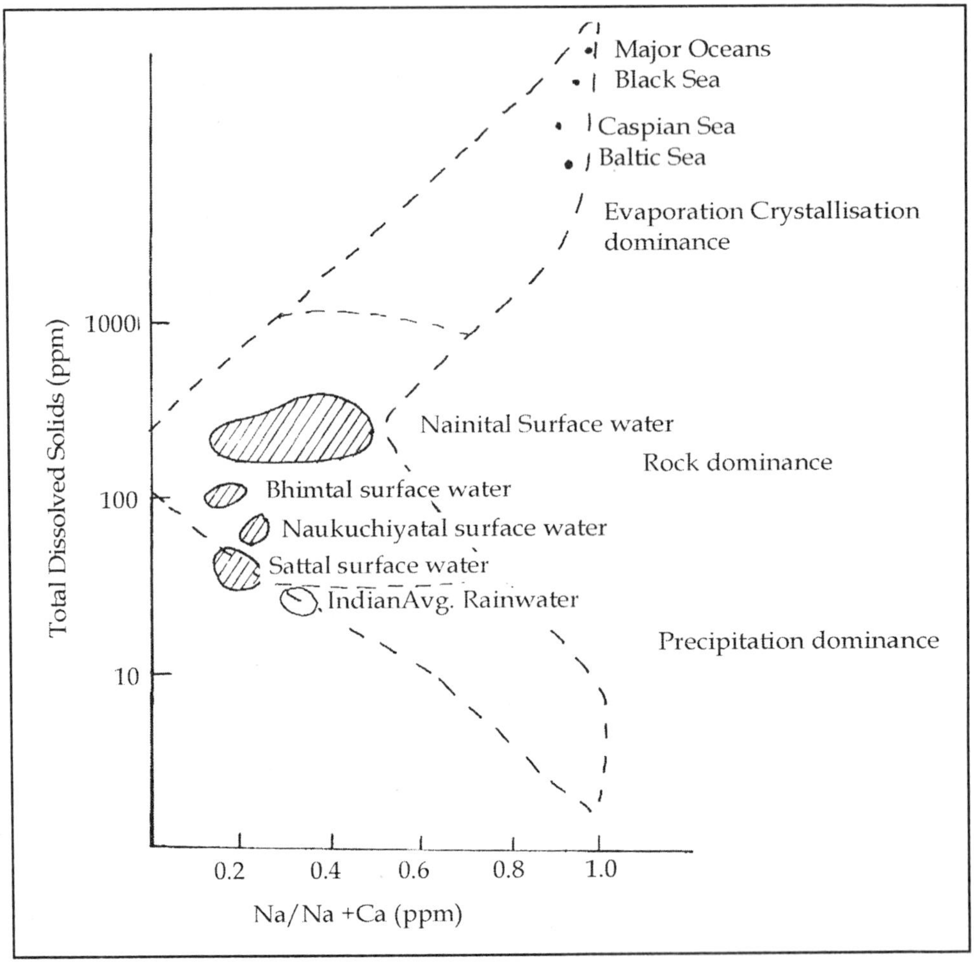

Fig. 5.31. The Gibb's (1970) diagram demonstrates that the Nainital Lake water is rock dominant.

The Nainital Lake water is rich in nutrients. The phosphate concentration is 124 μg/l and show increasing trend from 70 μg/l and 104 μg/l reported earlier (Das and Pandey, 1978; Handa *et al.*, 1982). Similary, there is increasing trend of ammonical nitrogen (NH_3 – N) from 19 mg/l in 1954 to 156 mg/l in 1975 and a significant increase in nitrite nitrogen (NO_2 – N) and nitrate nitrogen (NO_3 – N) from traces to 182 μg/l and 388 μg/l, respectively (Pant and Sharma, 1978; Pant *et al.*, 1980; Sharma *et al.*, 1982). This shows that the water quality of the lake is deteriorating due to increasing anthropogenic pollution in the lake, which is related to population increase in Nainital. A comparison of chemical data given by Handa *et al.* (1982) has shown that in the last decade also there has been increasing amount of most of the constituents in the lake water particularly PO_4, Na, Cl, and K indicating increasing anthropogenic flux in the lake which is attributed to sewage discharge, tourist influx, hotel construction etc.

Bhimtal Lake

In the vicinity of Nainital another beautiful lake Bhimtal exist which lies within geographical limits of 20°20' 19" N to 29°21' 00" N longitude and 79°33' 30" E to 79°34' 25" E latitude. The lake is a tourist attraction but due to anthropogenic activity the ecosystem of lake is adversely affected. In recent years, extensive road and house building activities on the hill slopes of the catchment area and increased industrialization have resulted in large scale land slides and consequent increase in sedimentation (4.70 mm/yr) in the lake (Das et al., 1994; Singh and Das, 1996).

Somewhat triangular in shape (Fig.5.32 a, b), with a length of 1.8 Km and width of 0.45 Km and catchment area of 4.5 sq km, Bhimtal Lake is located at an altitude of 1356 m above msl. The maximum depth of the lake is 26.5 m and is fed by a stream known as Nauli Godna flowing from the north-west through meta-basic (trap) rocks.

Fig. 5.32 (a). A panoramic view of Bhimtal Lake showing the catachment hill with poor vegetation due to volcanic and quartzitic rocks.

Fig. 5.32 (b). An island with a restaurant in the middle of the Bhimtal Lake.

Bhimtal is a tectonic lake and came into existence due to tectonic activity of Himalaya with development of fault and land slides blocking river flowing at that time and converting it into a lake during Holocene period. The average annual rainfall in the area is reported to be 200 cm/yr. The lithology around the lake comprises of metabasites associated with meta-sedimentary formations.

These predominantly consists of quartzits with grits, conglomerate with phyllite and are associated with prehnite-pumpellyite-metagreywacke and zeolite facies rocks changing to greenschist facies at places (Varadarajan, 1974) (Fig.5.33). Bhimtal Lake is alkaline in nature with average pH ranging from 8.54 to 9.10 indicating greater photosynthesis making water under saturated in CO_2 due to its assimilation faster than replenishment from the atmosphere which is a slow process (Stumm & Morgan, 1970; Holland, 1978) (Fig.5.34). Broad open Valley of Bhimtal compared to narrow valley of Nainital receive more sunlight causing greater photosynthesis. The E.C. of lake water is on an average 181 μs/cm which is lower indicating less ionic concentration (Table 5.16) (Das, 2005). The silicate lithology in the catchment which restrains greater dissolution of ion in the water compare to carbonate rocks explains the low ionic concentration of lake water.

The measured ionic strength of Bhimtal Lake is 3.5 x 10^{-3} mmoles/lt. The bicarbonate constitutes 62% (av) of the total anions in the lake water and sulphate 30.20%. Among cations, Ca constitutes 55.36% and Mg 31.0%

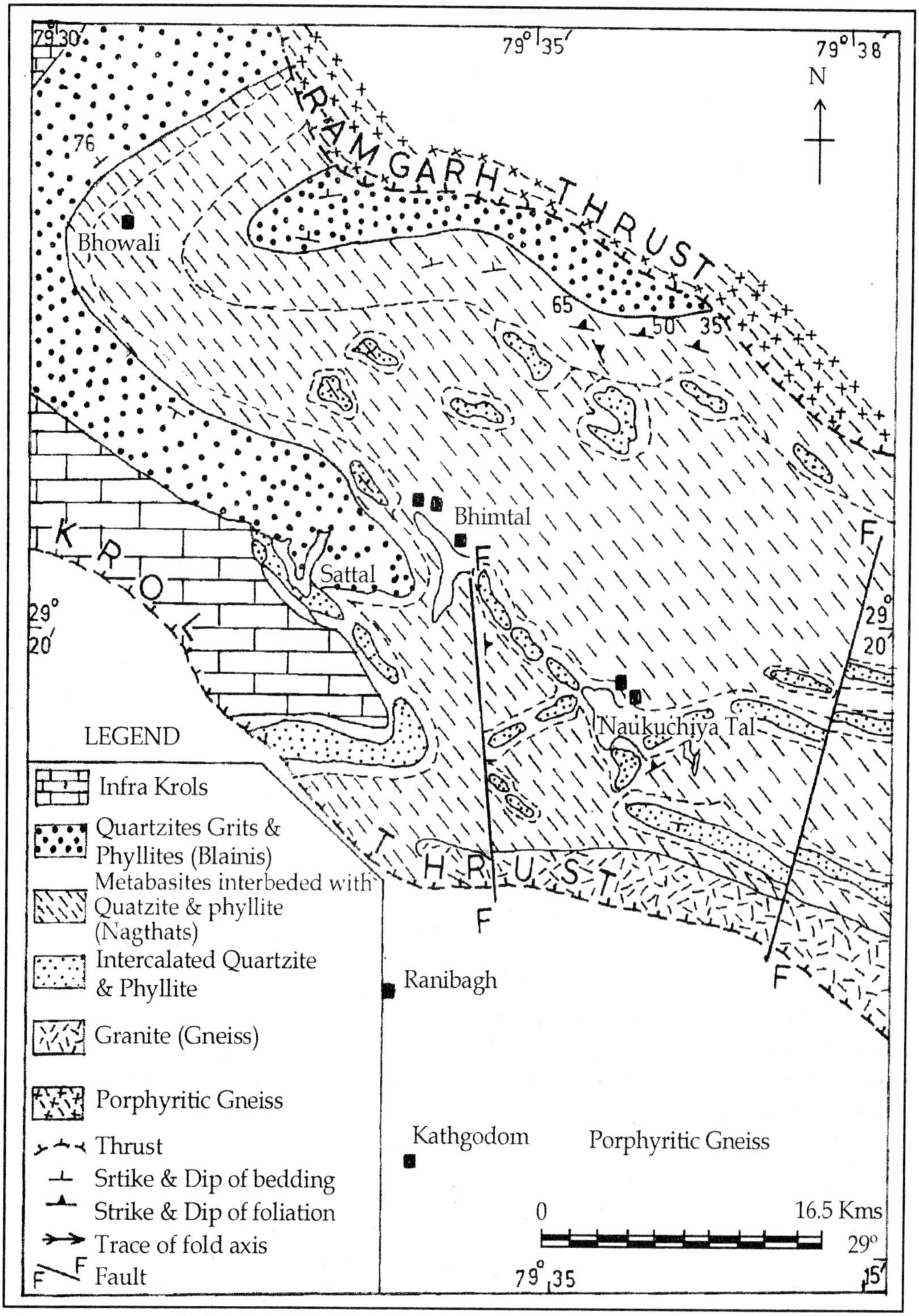

Fig. 5.33. Geological map of Bhimtal, Naukuchiyatal and Sattal Lakes showing lithology and structures (after Varadarajan, 1974).

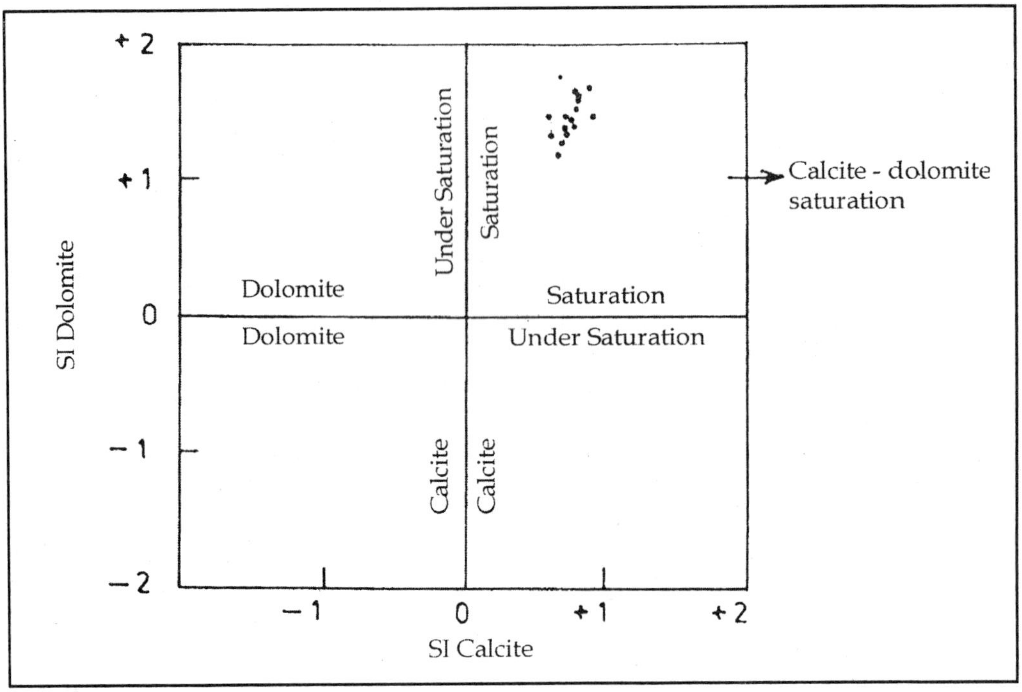

Fig.5.34. Saturation Index diagram showing that the Bhimtal Lake water is supersaturated with respect to calcite and dolomite.

(mean values) to the total cation budget of the water (Table 5.17). Although Ca and Mg together accounts for more than 80% of the total cation budget, the limestone does not occur in the catchment area. This shows that Ca-bearing minerals like amphibole, Ca-plagioclase occurring in metabasites and other Ca-minerals occurring in zeolite and prehanite-pumpyllite facies rocks on weathering release Ca and HCO_3 to the waters (Fig.5.31). The Ca^{2+} enrichment over Mg^{2+}, Na^+ over K^+ and low carbonate alkalinity 1.48 meq/l indicate water draining silicate rocks. It is an established fact that weathering derived bicarbonate comes from two sources, one is carbon in carbonate minerals such as calcite and dolomite and other arises as a result of the reaction of carbondioxide dissolved in soil water and groundwater with carbonate and silicate minerals. The carbondioxide is derived almost entirely from the bacterial decomposition of soil organic matter.

$$\underset{\text{Calcite}}{CaCO_3} + CO_2 + H_2O = Ca^{2+} + 2HCO_3^-$$

$$\underset{\text{Albite}}{2NaAl\,Si_3O_8} + 2CO_2 + 11H_2O \rightarrow \underset{\text{Kaolinite}}{Al_2Si_2O_5\,(OH)_4} + 2Na^+ + 2HCO_3 + 4H_4SiO_4$$

For both reactions weathering results in an accumulation of HCO_3^- ion in water. However, there is an important difference in the stoichiometry of the

Table 5.16: Chemical characteristics of Bhimtal Lake, Kumaun Himalaya

Sample No.	pH	E.C. (μs/cm)	DO	CO_3^{2-}	HCO_3^-	Cl^-	PO_4^{2-}	SO_4^{2-}	SiO_2 mg/l	Na^+	K^+	Ca^{2+}	Mg^{2+}	T.H	TDS	TSM
1.	8.84	180.2	7.3	6	122.5	6.39	0.0057	50	3.62	4.23	2.07	19.23	6.33	74.52	120	0.7
2.	8.85	182.0	7.4	6	87.5	9.94	0.0060	36	3.51	4.23	2.07	17.63	8.27	78.61	110	1.6
3.	9.10	177.4	8.2	12	87.5	5.68	0.0055	24	3.44	4.23	1.72	20.04	5.84	74.50	110	0.3
4.	8.54	190.3	8.7	00	87.5	7.81	0.0060	50	3.69	4.50	2.90	20.04	6.57	77.55	110	1.0
5.	8.82	179.3	8.4	18	90	5.68	0.0060	63	3.50	4.23	2.07	20.44	5.60	73.50	100	1.2
6.	8.94	183.8	8.4	24	88.5	5.68	0.0045	43	3.85	4.76	2.07	20.44	5.60	73.51	110	2.5
7.	8.87	179.4	8.6	12	86.5	7.10	0.0080	30	3.86	4.23	1.74	17.63	10.22	86.74	110	1.3
8.	8.92	182.0	9.6	00	90	6.39	0.0110	76	3.92	4.50	2.07	18.43	7.30	77.25	110	5.9
9.	9.04	191.1	8.3	00	86.5	6.39	0.0078	24	3.88	4.50	2.07	22.44	7.30	87.40	110	6.5
10.	8.97	182.2	8.6	24	102	7.10	0.0068	16.5	3.72	4.50	2.07	20.04	6.57	78.15	110	1.2
11.	8.93	182.7	8.4	24	95	5.68	0.0055	39	3.65	4.50	2.07	19.23	7.30	79.29	110	6.1
12.	8.81	182.6	8.1	00	86.5	6.39	0.0055	43	4.00	4.23	2.07	21.04	5.84	77.69	110	3.9
13.	8.85	179.5	10.1	00	86.5	4.97	0.0068	43	3.69	4.50	2.07	19.63	6.08	75.14	110	1.1
14.	8.90	177.5	9.6	12	87.5	6.39	0.0070	39	3.64	4.23	2.07	18.43	7.30	77.25	110	0.4
15.	8.95	175.4	10.0	39	80	7.10	0.0115	30	3.88	4.50	2.07	19.23	5.84	73.12	100	1.2
16.	8.86	182.5	8.6	6	95	6.39	0.0057	33	3.68	4.50	2.07	22.84	6.08	83.25	110	5.3
17.	9.00	179.5	9.9	12	90	6.39	0.0060	36	3.59	4.23	2.07	18.03	7.30	76.74	110	1.2
18.	9.00	174.6	8.1	30	91.5	5.68	0.0057	27	3.84	4.23	2.07	23.24	3.89	75.04	100	0.8
19.	8.96	181.3	11.2	24	95	4.97	0.0080	19	5.90	4.50	2.07	18.43	8.27	81.33	110	4.0
20.	8.92	175.2	11.0	00	86.5	5.68	0.0060	19	3.69	4.23	2.07	19.63	6.33	76.20	100	0.8

Table 5.17: Percentage of Equivalent per million of the constituents in Bhimtal Lake

Sample No.	..............Cations...............			..............Anions...........		
	Ca	Mg	(Na+K)	(CO_3+HCO_3)	SO_4	Cl
1.	55.72	30.56	13.72	62.19	32.25	5.56
2.	48.76	38.12	13.12	58.21	30.43	11.36
3.	58.37	28.36	13.28	68.48	23.87	7.65
4.	55.08	30.09	14.83	53.19	38.65	8.16
5.	59.22	27.04	13.47	50.13	44.45	5.42
6.	58.48	26.66	14.86	57.86	35.76	6.38
7.	44.93	43.43	11.64	63.21	27.88	8.91
8.	51.83	34.20	13.97	45.55	48.88	5.57
9.	56.69	30.73	12.58	67.58	23.85	8.57
10.	55.74	30.44	13.82	75.48	15.50	9.02
11.	52.87	33.45	13.68	61.55	32.12	6.33
12.	59.26	27.40	13.34	56.86	35.92	7.22
13.	56.51	29.17	14.32	57.78	36.52	5.70
14.	52.15	34.43	13.42	59.09	33.48	7.43
15.	56.68	28.67	14.65	61.38	29.26	9.36
16.	60.19	26.70	13.11	64.23	28.35	7.42
17.	51.62	34.82	13.56	66.96	22.62	6.42
18.	67.44	18.82	13.74	67.49	25.32	7.19
19.	49.56	37.06	13.38	74.39	18.93	6.68
20.	56.21	30.21	13.58	71.84	20.05	8.11

two types of reactions. For weathering of carbonate minerals like calcite, dolomite and aragonite the HCO_3^- ion concentration is balanced by cations like Ca^{2+}, Mg^{2+} on an equivalent basis. On the other hand, kaolinisation of silicate minerals results in SiO_2 in solution which is related to the HCO_3^- ion concentration through the composition of the primary silicate (Garrels, 1967). In Bhimtal Lake, Ca + Mg equivalent ratio is 0.52 which is not in conformity with stoichiometric requirements hence does not support weathering of carbonate minerals releasing Ca, Mg and HCO_3^-. The low Ca + Mg/Tz^+ ratio (0.86), low (Ca +Mg)/(Na + K) equivalent ratio 3.26 and weak (Na* + K)/Tz^+ ratio (0.13) suggest that elemental ions are derived from Ca-bearing silicate phases, present in the rock formations outcropping in the lake basin and secondary calcite present in voids or vein fillings of metabasite (Varadharajan, 1974). Due to high dissolution of calcite, Ca, Mg and HCO_3^- predominates in water. The low phosphate content (7µg/l) in Bhimtal Lake indicates low anthropogenic input compared to Nainital. The dissolved silica content ranges between 3.44 to 5.90 mg/l and is higher than in Nainital. The higher dissolved silica is due to silicate lithology in the catchment and also due to lower phosphate loading which controls silica consuming diatoms population in shallow waters of the lake (Schelske, 1988). The HCO_3/SiO_2 equivalent ratio in this lake is 23.74.. This shows silicate contribution because in Nainital this ratio is high 106.2 as the HCO_3^- concentration is derived from carbonate lithology.

Naukuchiyatal Lake

Naukuchiyatal Lake is situated at an altitude of 1219 m above msl. The lake is 0.95 km in length and 0.69 km in width having maximum depth of 40.2 m, a deepest lake in the Nainital group. The catchment area is about 2.8 sq. km and the hills surrounding the lake are quite densely vegetated with very sparce settlements and population (Fig.5.35 a,b).The pristine character of the lake is well protected. The Naukuchiyatal and Bhimtal lakes are almost in the same (NW-SE) alignment, which is the trend of the hill ranges following the lake catchment. Massive landslide blocks are seen on the northern edge of the Naukuchiyatal and a number of terraces in between the landslide masses indicate that Bhimtal and Naukuchiyatal were perhaps one single lake, which got detached due to tectonic upliftment followed by landslides.

Fig. 5.35 (a). Naukuchiyatal – a beauty in wilderness.

Fig. 5.35 (b). Naukuchiyatal Lake showing thick vegetation and sparce population. The lake is unpolluted .

The rock formation comprising the lake catchment are volcanic rocks of greenish grey colour, fine grained and interbedded with quartzites. The vesicles are often filled with secondary minerals like quartz, epidote, chlorite, zeolite and calcite. The volcanics are penetrated by dykes and sills of relatively fresh and slightly metamorphosed dolerites and are younger than host rocks.

The water chemistry of Naukuchiyatal Lake is characterized by low carbonate alkalinity of 1.22 meq/l on average, low Ca abundance between 12.0 to 16.0 mg/l that contributes on an average 52% of total cation, Mg between 4.87 mg/l to 7.30 mg/l and contributes 32.5% of the cationic budget. The bicarbonate ranges from 70 mg/l to 80 mg/l and constitutes 70.45% of the total anions budget of lake water (Table 5.18 and 5.19). The low alkalinity, low Ca^{2+} and Mg^{2+} and low HCO_3^- concentration in lake water do not indicate limestone dissolution and is in conformity with the exposed lithology. The high silica concentration ranging from 5.88 to 7.05 mg/l and (Na +K)/Tz^+ equivalent ratio of 0.16 and low (Ca +Mg)/ Tz^+ 0.84 equivalent ratio, indicates silicate and volcanic rock weathering releasing cations and anions to the lake water (Fig.5.31). Therefore, Ca^{2+} concentration in lake water is derived from Ca-bearing silicate phases in metabasic rocks and secondary Ca- minerals filling the vesicles and veins in volcanic rocks particularly calcite, epidote etc. A positive correlation between pH and HCO_3^- and very low positive correlation

Table 5.18: Chemical characteristics of Naukuchiyatal, Kumaun Himalaya

Sample No.	pH	E.C. (μs/cm)	DO	CO_3^{2-}	HCO_3^-	Cl^-	PO_4^{2-}	SO_4^{2-}	SiO_2	Na^+	K^+	Ca^{2+}	Mg^{2+}	T.H	TDS	TSM
			mg/l													
1.	9.60	145.7	9.6	nil	80.00	8.52	0.0055	10.0	5.88	4.50	1.03	15.23	5.84	63.02	80	0.3
2.	9.50	150.3	9.6	nil	70.00	7.10	0.0078	13.0	6.44	4.23	1.37	15.23	4.87	58.94	80	2.8
3.	9.44	150.3	9.4	nil	75.00	7.10	0.0068	16.5	6.30	4.50	1.37	16.03	4.87	60.96	80	1.5
4.	9.49	149.2	9.5	nil	75.00	7.10	0.0055	16.5	6.80	4.23	1.03	15.23	4.87	58.92	80	0.8
5.	9.46	146.6	9.7	nil	75.00	5.68	0.0078	3.0	6.28	4.23	1.37	15.23	5.36	61.00	80	0.7
6.	9.58	141.9	9.6	nil	70.00	7.10	0.0060	4.0	6.40	4.23	1.03	14.44	5.36	59.01	80	0.9
7.	9.51	145.5	9.8	nil	80.00	6.39	0.0060	13.0	6.20	4.23	1.03	14.42	5.84	60.98	80	1.9
8.	9.46	146.5	9.6	nil	75.00	7.10	0.0057	16.5	6.19	4.23	1.03	12.02	7.30	61.06	80	1.1
9.	8.51	151.3	9.7	nil	70.00	5.68	0.0070	19.0	7.05	4.50	1.37	14.44	5.84	61.03	90	0.7
10.	9.41	147.3	9.6	nil	70.00	7.10	0.0100	33.0	6.04	3.96	1.37	14.42	4.87	56.90	80	0.9
11.	9.46	148.5	9.1	nil	75.00	5.68	0.0045	10.0	6.21	4.23	1.30	14.42	5.36	58.96	80	1.0

Table 5.19: Percentage of Equivalent per million of the constituents in Naukuchiyatal Lake

Sample No.	..............Cations...............			..............Anions............		
	Ca	Mg	(Na+K)	(CO_3+HCO_3)	SO_4	Cl
1.	51.79	33.09	15.12	74.51	11.84	13.65
2.	54.93	29.28	15.79	70.80	16.78	12.42
3.	55.73	28.23	16.04	69.37	19.35	11.28
4.	55.29	29.45	15.26	69.37	19.35	11.28
5.	53.35	31.29	15.36	84.71	4.28	11.01
6.	52.28	32.39	14.73	61.95	27.19	10.86
7.	50.18	34.33	14.83	74.43	15.45	10.22
8.	42.33	42.85	14.82	69.37	19.35	11.28
9.	58.18	33.80	16.02	67.25	23.32	9.43
10.	54.04	30.43	15.53	56.24	33.90	9.86
11.	52.07	32.25	15.68	76.97	13.02	10.01

between pH and Ca^{2+} also suggests poor dissolution of $CaCO_3$ due to lack of carbonate rocks in the area and deficiency of CO_2 resulting in high pH, low HCO_3^- and Ca^{2+} in the lake waters (Table 5.18 and 5.19). The lake water is deficient in sulphate concentration and it contributes 18.53% to the total anion budget. This shows that there is no significant source of SO_4^{2-} in Naukuchiyatal Lake basin. A negative correlation of SO_4^{2-} with Ca^{2+} indicates absence of gypsum in the source rock. The Na and K content in lake water are also low and both together contribute 15.38% of the total cation budget.

The sandstone is unlikely to contain sodic phases except for albite (Na-plagioclase) and therefore source of soda is from other sources such as amphiboles in the metabasic rocks or in secondary zeolites in volcanic rocks. Very low phosphate concentration (6.6μg/l) indicates insignificant anthropogenic interference and the lake is oligotrophic (Fig.5.35) maintaining its pristine status. Since the lake is not polluted, as is evident from low phosphate content, which promotes growth of phytoplankton and diatoms etc., therefore, depletion in silica due to uptake by these micro-organisms is unlikely and dissolved silica content in water is high 6.34 mg/l or 0.105 mole/l on an average. The HCO_3^-/SiO_2 average value is 11.52 on equivalent basis, which also indicates silicate source. The lake shows a normal trend Ca> Mg > Na > K and HCO_3>SO_4>Cl and water represents Ca-HCO_3 type of facies (Fig.5.36).

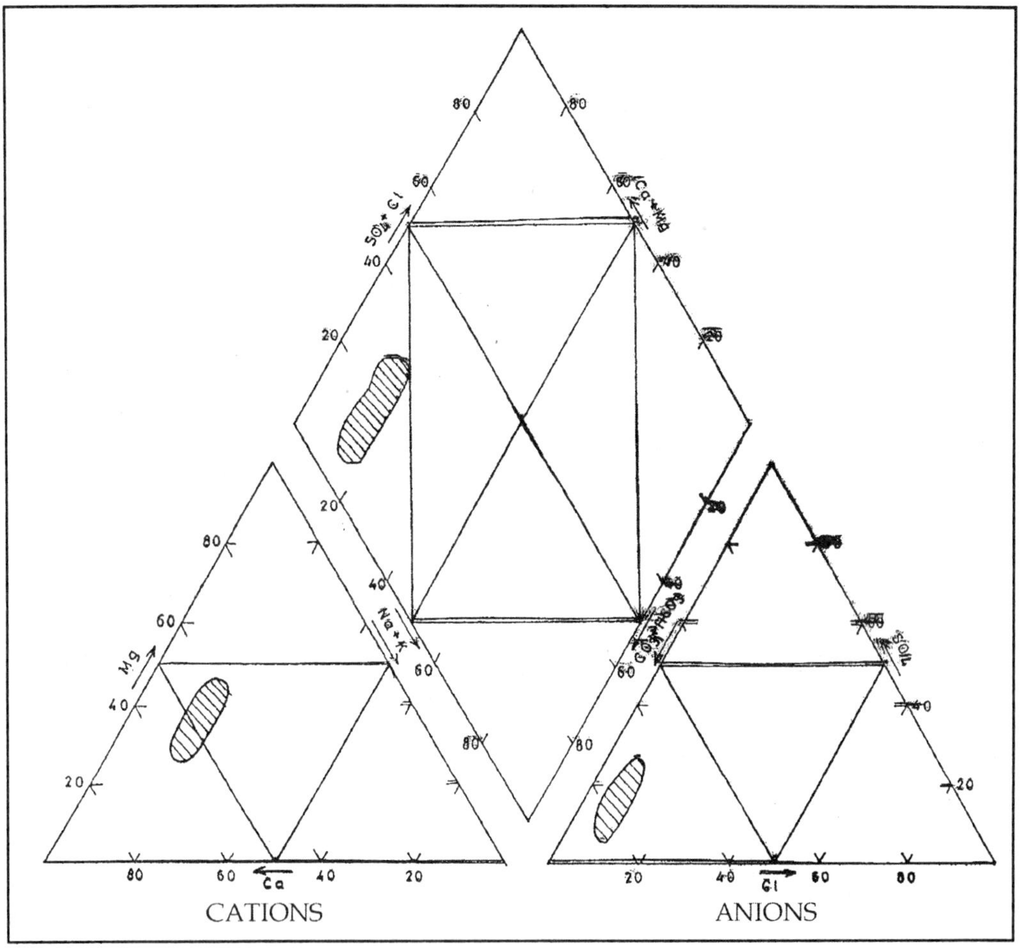

Fig. 5.36. In a trilinear diagram Naukuchiyatal Lake water distinguishes as Ca - HCO_3- SO_4 type.

Sattal Lake

Another lake of Nainital group is Sattal at an altitude of 1286 m above msl. It is 1.3 km in length and 0.2 km in width with a maximum depth of 20 m and a minimum of 8 m having a catchment area of 428 sq km. The name Sattal symbolizes seven Tals (ponds) flowing from different directions and meeting at one point and having different mythological names thus acquiring a peculiar shape (Fig.5.37 a, b). It is V-shaped with two prominent arms, the western arm has a contact with a almost N-S running fault (Sattal Fault) and the landmass to the east of the fault subsided resulting in the damming up of the stream forming lake during Holocene period. Seven lakes or depressions of the area were formed as a result of debris cones blocking paths of ravines. The lake is situated on Bhowali quartzite and its southern bend is in contact with limestone and eastern in contact with dolerite (Fig. 2.3).

Fig. 5.37 (a). Sattal – a virgin beauty in wilderness.

Fig. 5.37 (b). Dense vegetation on the hills surrounding Sattal Lake.

The Sattal Lake is alkaline with pH ranging from 9.53 to 9.72 with an average of 9.66. The E.C and TDS are low 119 μs/cm and 62 mg/l, respectively showing low ionic concentration (2.0 x 10^{-3}) in waters (Table 5.20). The low partial pressure of CO_2 ($pCO_2^{10-4.82}$) compared to ambient log $pCO_2^{-3.5}$ reflects escape of CO_2 from the lake water or it is consumed during photosynthesis increasing the pH of the water and precipitation of Ca^{2+} which is evident from the saturation indices plot (Fig.5.38) in which water samples lie in supersaturated zone with respect to both calcite and dolomite. The low content of Ca^{2+} in water is probably due to depletion of Ca in dissolved form. However, in a weathering of carbonate minerals like calcite and dolomite, the HCO_3^- ion concentration is balanced by cations like Ca^{2+}, Mg^{2+} on an equivalent basis which is not precisely the case in this lake as, $Ca^{2+} + Mg^{2+}$: HCO_3^- equivalent ratio is 1.14.

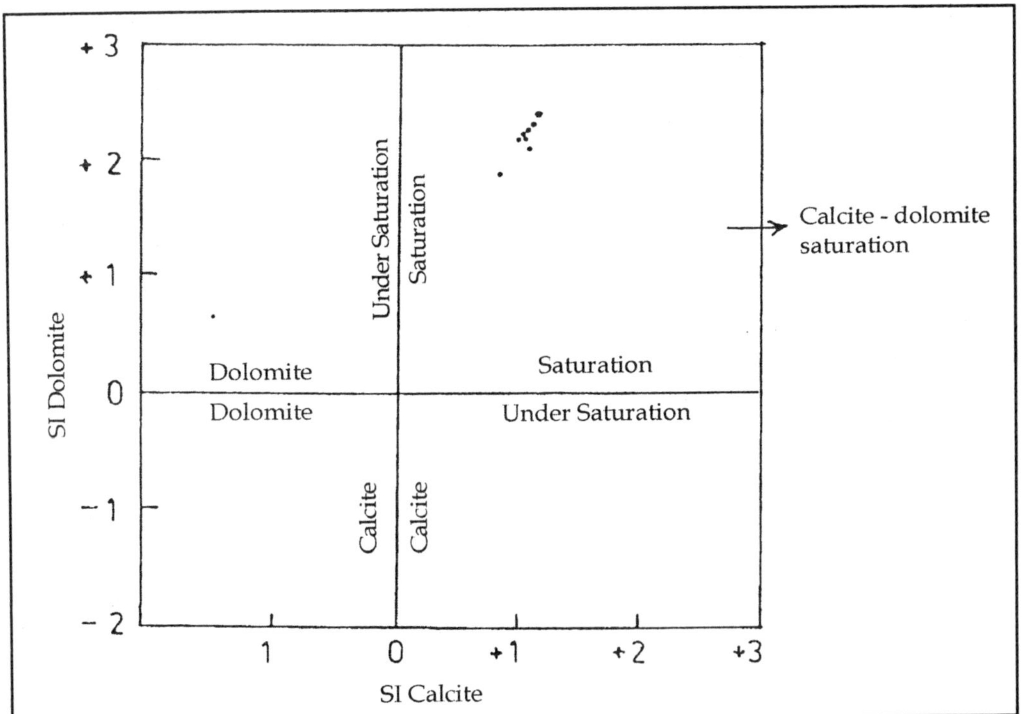

Fig. 5.38. In a saturation indices plot Sattal surface water falls in over saturated domain with respect to calcite and dolomite.

The (Ca +Mg)/Tz^+ value which is generally considered to indicate carbonate and silicate rock weathering is 0.88 and is less than 0.9. This is supported from Na* + K/Tz^+ ratio of 0.19 indicating dissolution of both carbonate and silicate phases in the watershed. The Ca^{2+}/HCO_3^- and $Ca^{2+}+Mg^{2+}/HCO_3^-$ molar ratio are 0.31 and 0.56 respectively, which are also not the same 0.50 as reported by Zhang and others (1995) in case of carbonate

Table 5.20: Chemical characteristics of Sattal Lake, Kumaun Himalaya

Sample No.	pH	E.C. (μs/cm)	DO	CO_3^{2-}	HCO_3^-	Cl^-	PO_4^{2-}	SO_4^{2-}	SiO_2	Na^+	K^+	Ca^{2+}	Mg^{2+}	T.H	TDS	TSM
			mg/l													
1.	9.67	123.7	9.1	nil	50.00	5.68	0.0085	16.5	5.45	2.64	0.69	12.02	5.36	52.90	70	1.7
2.	9.71	122.7	11.0	nil	58.00	8.52	0.0070	10	5.40	2.64	0.69	11.22	5.36	50.88	70	4.0
3.	9.63	118.6	10.2	nil	58.00	9.94	0.0057	13	5.78	2.64	0.69	10.42	5.84	50.88	60	5.3
4.	9.66	118.5	10.6	nil	58.00	7.10	0.0060	16.5	5.64	2.37	0.69	11.22	5.36	52.90	60	1.2
5.	9.72	116.8	10.1	nil	50.00	8.52	0.0080	67	5.32	2.64	0.69	11.22	5.36	52.90	60	4.2
6.	9.65	117.5	9.8	nil	58.00	6.39	0.0070	24	5.62	2.90	0.69	11.22	5.84	52.90	60	1.4
7.	9.70	119.6	10.4	nil	50.00	7.10	0.0080	19	6.70	2.90	0.69	12.02	4.38	48.78	60	2.4
8.	9.53	115.8	10.2	nil	50.00	5.68	0.0068	3	5.16	2.90	1.03	9.61	5.84	48.83	60	3.4
9.	9.68	119.7	9.9	nil	58.00	7.10	0.0070	6	5.30	2.64	0.69	11.22	5.36	50.88	60	2.3

weathering. The molar ratio Ca/Na ~2.40, Mg/Na ~ 1.93, HCO_3/Na ~7.65 are lower than those of the waters that drain carbonate lithology and are higher than those that drains silicate rocks (Stallard, 1980; Meybeck, 1987; Negrel and others, 1993). Therefore, these ratios indicate that carbonate and silicate end members are the major contributors of major elements to the water chemistry of Sattal Lake. The Ca/Na molar ratio of the average crustal sedimentary rocks is close to 0.6 (Taylor and McLennan, 1985) and, due to the higher solubility of Na to Ca, lower molar ratios are expected in the dissolved loads of rivers draining silicate rocks. Thus, a higher ratio of Ca/Na in Sattal Lake indicates non-silicate source, which contribute Ca in the basin. The chemical data of lake water therefore, favour both silicate and carbonate rock weathering in the watershed and calcite and Ca-bearing minerals being readily soluble in water, enhances its content compared to other ions. This is consistent with the silicate, metabasic and carbonate rocks lithology outcropping in the area.

In Sattal CO_3^{2-} + HCO_3^- contribute 62.18% whereas SO_4^{2-} 23.66% of total anion budget and the correlation between Ca^{2+} and SO_4^{2-} is positive (r=0.17), suggesting a common source (Table 5.21). Na^+ + K^+ contribute on an average 11.83% to the total cation budget of the lake and the dominance of Na^+ over K^+ in this fresh water lake is due to greater mobility of Na^+ during weathering. K^+, being less mobile and less soluble, remains in the residue. It is envisaged that the depletion in K^+ is due to non-availability of K-bearing phases such as mica, K-feldspar etc. in the lithology of the area. This is reflected from the clay mineral study by X-ray diffraction, which reveals a high content of chlorite and very much less illite in the sediments.

Table 5.21: Percentage of Equivalent per million of the constituents in Sattal Lake

Sample No.	...............Cations...............			...............Anions...........		
	Ca	Mg	(Na+K)	(CO_3+HCO_3)	SO_4	Cl
1.	50.93	37.85	11.22	61.98	25.93	12.09
2.	50.94	37.85	11.21	67.95	14.89	17.16
3.	45.72	42.68	11.60	63.33	18.01	18.66
4.	49.72	39.60	10.68	63.60	23.01	13.39
5.	49.21	39.21	11.58	33.47	56.74	9.79
6.	47.09	40.84	12.03	58.28	30.68	11.04
7.	54.16	32.89	12.95	57.90	27.98	14.12
8.	42 92	43.46	13.62	78.69	5.96	15.35
9.	49.21	39.17	11.62	74.50	9.82	15.68

The dissolved SiO_2 varies from 5.16 to 6.70 mg/l (0.093 mmol/l average). According to Meybeck (1980) in arctic rivers (average temperature <4°c) SiO_2 is 3 ± 2 mg/l, temperate rivers (average temperature 4° to 19°c) have an average SiO_2 concentration of about 6 ± 3 mg/l and tropical rivers (average temperature > 20°c) contain about 13 ± 4 mg/l SiO_2. He attributes the temperature effect on dissolved silica to the differences in silicate weathering products (clays) formed in different climatic zones. In the tropics, chemical weathering is more complete and silicate minerals tend to weather to kaolinite, which releases 1.5 times as much dissolved silica as weathering to smectite, which is more common in temperate areas. As the climate of this region is cool with good rainfall, silica released during weathering is high. The HCO_3^-/SiO_2 molar ratio 9.61 also supports quartzite, meta-basites, and limestone controlling the water chemistry of this lake. In the absence of major anthropogenic interference, the study is more revealing. The phosphate content in this lake is 7.10 µg/l, which indicates low nutrient loading and low pollution hence the lake water can be used for agriculture or even for potable purposes.

Lakes of Udaipur (Rajasthan)

Fateh Sagar Lake

Udaipur is a Saucer-shaped basin known as the Ahar valley and is drained by the north-west flowing River Ahar and its tributaries. Although rainfall is very uncertain, being a desert region, an average annual rainfall of 635 mm has been recorded. The lakes are about 587 m above msl. Fateh Sagar Lake is somewhat pear shaped (Fig.5.39) with a length of 2.6 km and width of 1.8 km; it is surrounded by hills of Aravalli Mountains. The main feeder canal to this lake comes from Madar tank located about 15 km from Udaipur city. The maximum depth of the lake is 13.4 m and the mean depth is 5.4 m. The rate of sedimentation is 2.42 mm/yr (Das and Singh,1994). The catchment lithology of Fateh Sagar Lake consists predominantly of phyllite and greywacke belonging to Precambarian Aravalli Group rocks. Phyllite often contains pyrite cubes embedded in it.

The water of the lake is alkaline and pH ranges from 8.91 to 9.23 with an average of 9.12. The cause of high pH has been considered in terms of natural processes of weathering, evaporation ,and CO_2 gas equilibration (Garrels and Mackenzie, 1967). The HCO_3^- average value is 255.69 mg/l and Ca and Mg is 17.13 and 23.10 mg/l on an average, respectively (Table 5.22).

The HCO_3^- value is more than twice the concentration of Ca^{2+} +Mg^{2+}. The weathering of feldspars, result in the production of groundwater containing dissolved HCO_3^- that is balanced by Na^+ and K^+ as well as Ca^{2+} and Mg^{2+}. Upon evaporation of the groundwater after passing in the lake, the lake cannot precipitate all HCO_3^- as Ca^{2+} or Mg^{2+} carbonates and some HCO_3^- remains behind to be concentrated by evaporation. The chemical weathering of sedimentary silicate rocks is further supported from the computation of HCO_3^- derived from carbonate and silicate source rocks. In case of the former, the average value in milliequivalent per litre is 1.403, whereas in the latter it is 2.787which is about two times more than the carbonate minerals.

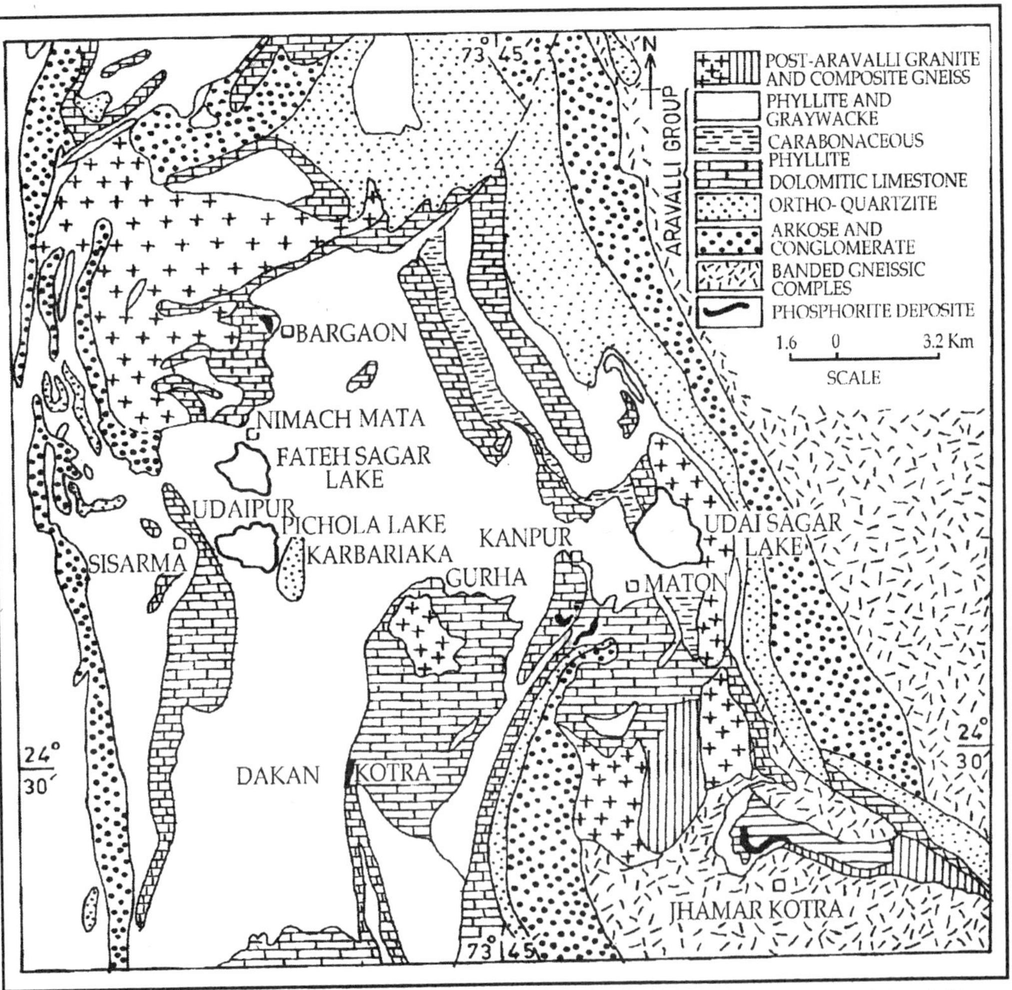

Fig. 5.39 Geological map of Udaipur showing configuration and lithology of Feteh Sagar Lake.

The concentration of Na is 61.03 and K is 3.94 mg/l, which are more than 1.5 times Ca + Mg combined together. The Cl concentration of 58.22 mg/l is also close to that of Na. In the absence of salt deposit in the area, and the marine aerosol fall out, the latter has been ruled out which finds support from isotopic studies of the Sambhar Lake, located at the eastern fringe of the Thar desert in western India (Ramesh et al.,1993; Yadav, 1997). Therefore, the source of soda and chloride is presumably from sodic phases present in the rock, anthropogenic sources, and evaporation. The sedimentary rocks

Table: 5.22: Chemical characteristics of Fateh Sagar Lake

	pH	E.C. (µs/cm)	DO	CO_3	HCO_3	Cl	SO_4	PO_4	Ca	Mg	Na	K	SiO_2	TDS
Site			amount meq/l						cation (%)			anion (%)		
FS1	8.91	5.77	7.5	12	244.0	61.06	66	0.097	16.83	26.29	61.80	3.91	1.60	360
FS2	9.08	5.68	8.1	12	256.2	59.64	27	0.094	16.42	20. 21	74.15	3.91	2.60	360
FS3	9.23	563	9.0	15	244.0	58.22	106	0.090	19.23	20.94	71.05	3.68	2.48	360
FS4	9.21	569	11.0	9	262.3	59.64	50	0.090	16.88	23.37	58.70	3.91	2.77	360
FS5	9.21	563	9.0	12	231.8	56.80	-	0.094	16.02	23.37	68.00	3.68	3.15	360
FS6	9.12	566	9.8	15	256.2	53.96	150	0.088	16.83	23.37	55.60	3.68	2.72	360
FS7	9.17	466	9.4	12	268.4	58.22	1.60	0.090	16.83	27.75	52.55	3.68	2.95	3.60
FS8	9.13	559	7.5	9	262.3	59.64	42	0.090	16.83	22.88	55.60	3.91	2.68	350
FS9	9.09	560	8.5	9	250.1	59.64	63	0.086	16.83	22.40	56.65	3.91	2.70	350
FS10	9.14	558	9.3	12	231.8	56.80	50	0.094	17.63	21.91	55.60	3.91	2.77	350
FS11	9.19	557	9.0	12	298.9	58.22	53	0.094	16.83	23.37	74.15	4.56	2.68	360
FS12	8.96	557	8.8	12	262.3	56.80	76	0.088	18.43	21.42	62.55	4.56	2.77	360

contain Na-plagioclase $NaAlSi_3O_8$ (albite), which may be a major source of Na in ground, river, and lake water. However, Berner and Berner (1987) pointed out that considering all sources of Na, 28% is from pollution, 8% from sea salt, 42% from halite, and remaining 22% from weathering of plagioclases. According to this data, the silicate weathering and pollution will contribute 50% of the Na in the river and lake waters.

In many saline lakes evaporation path proposed by Hardie-Eugster (1970) has been followed according to which in most natural water evaporative concentration saturates calcium carbonate which precipitates and alkalinity is two times more (alk =4.58 meq/l) than Ca (Ca =0.567) and also magnesium silicate (mineral sepiolite) precipitates and alkalinity is two times more than Mg^{2+} (Mg =1.926 meq/l) which concentrates Na^+, SO_4^{2-} and Cl^-. The following reaction takes place:

$$2Mg^{2+} + 4HCO_3^- + 3H_4SiO_4 \rightarrow Mg_2Si_3O_8\ n\,H_2O + 4CO_2 + 8H_2O$$

In Fateh Sagar Lake, high evaporation particularly during high summer temperature, poor in flow and poor rainfall causes cyclic wetting and drying which aids in the formation of alkaline/saline soil and consequent saline groundwater and lake waters.

The high Cl content in the lake water may probably be due to high evaporation and cyclic wetting and pollution. Chloride is also added to surface water by redissolution of saline evaporative crusts in desert basins. Meybeck (1979) estimated that about 30% of Cl^- in river water arises from pollution. Domestic sewage contains considerable Cl^-, seepage from septic tanks, fertilizer, and industrial Cl^- containing brines etc. In Fateh Sagar Lake Cl^- is derived from the alkaline/saline soil under the process of cycling wetting and drying and from anthropogenic pollution.

The low K^+ content is due to its low mobility compared to Na and Ca. Meybeck (1984) estimates that in about three-fourths of silicate weathering, K comes from sedimentary silicates and one-fourth from igneous and metamorphic silicates. Holland (1978) estimates that on an average only about 50% of rock potassium is released in solution during silicate weathering. As the SiO_2 is also less (0.04 eq.value), it is likely that K has been used in the formation of silicate minerals like illite which constitute large percentage amongst clay minerals in sediment. The sulphate concentration in this lake is on an average 70.25 mg/l and is much higher than in Pichola Lake which shows concentration of 35.64 mg/l. The high value is due to oxidation of pyrite yielding high sulphate ions to the water body. However, the high pH of the lake water does not indicate H_2SO_4 formation. The phosphate content is 91 μg/l on an average and is similar to Pichola Lake which contains 97μg/l. This shows that both the lakes located in the Udaipur town are receiving anthropogenic flux and eventually becoming eutrophic. The SiO_2

concentration is low and HCO_3^-/SiO_2 molar ratio is high (64.28 meq/l). This indicates low SiO_2 dissolution from sandstone (greywack) rocks comprising mostly of quartz. Stallard (1980) has mentioned that silica concentration in igneous and metamorphic rivers are greater than in sedimentary silicate rivers of the Amazon Basin. This is because sedimentary silicates (sandstone and shale) release less silica as they consist largely of detrital minerals that are basically resistant to weathering (quartz, micas, and clays).

Pichola Lake

Pichola Lake in Udaipur district, Rajasthan, is about 3.62 km in length from north to south and 2.41 km in width from east to west. It has a mean depth of 5.6 m and covers an area of 9.71 sq km and contains estimated amount of 418 million cubic feet of water. It is mostly fed from rainwater and Sisarama tributary flowing from the west. In the north-east, there is a city palace of Maharaja whereas in the south-east Eklinggarh hill flanks the lake (Fig 5.40). The old town, with its ghats for washing and bathing is located at the northern end of the lake. About 30,000 inhabitants reside around the lake. An average annual rainfall of about of 73 cm has been recorded.

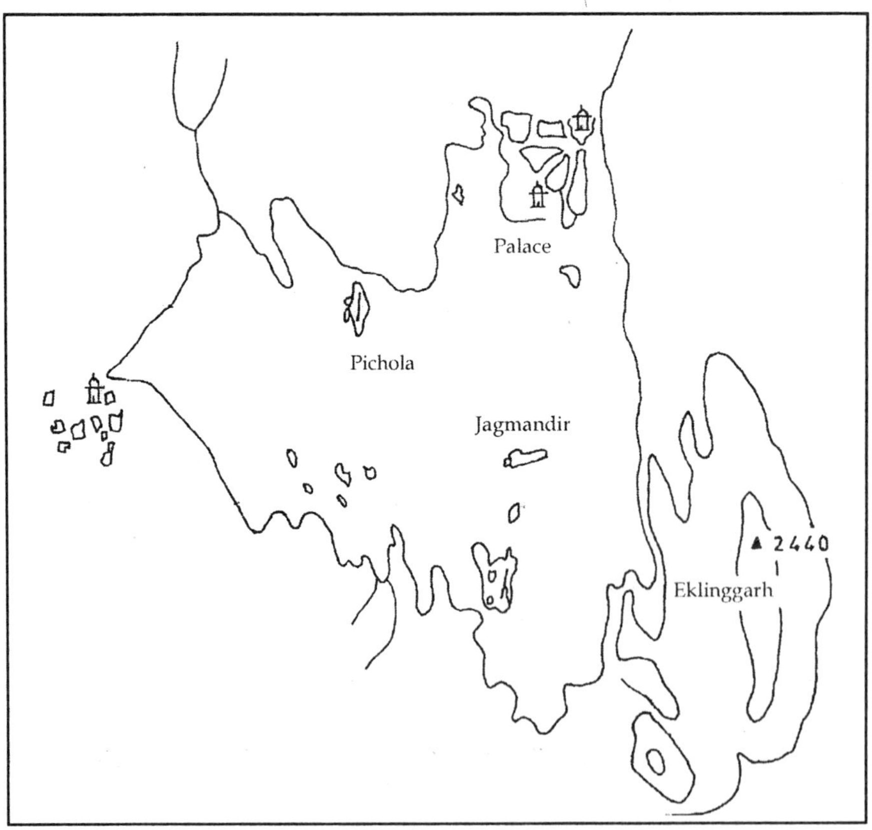

Fig. 5.40. Configuration of Pichola Lake showing an island over which Jagmandir Temple is built up and Palace is towards the north.

The hills around Udaipur are characterized by a great node of spurs and curving ridges exposing the highly denuded oldest Precambrian lithology of the Indian sub-continent. The main rock types of the catchment of Pichola Lake consists of slaty-quartzite, phyllite, greywacke often interlayered with massive orthoquartzites showing ripple marks, current bedding, and other sedimentary structures. This indicates shallow marine deposition and dolomitic limestone towards the west of the lake belonging to the Aravalli Super Group (Heron, 1953; Roy, 1988) (Fig.5.39).

Pichola Lake is alkaline with pH ranging from 8.84 to 9.27 with an average of 9.12 and is comparable to other arid climate lakes of India and the world. The high evaporation, poor flushing and low precipitation may allow higher residence time of water in the basin thereby increasing salinity. The HCO_3^- on an average is 235.42 mg/l, Cl 72.72, Na 73.27, Ca 21.28, Mg 21.36 mg/l have been measured (Table 5.23), (Das and Singh, 1996). The high concentration of Cl^- and Na^+ are not derived from marine aerosol as discussed above. The Pichola Lake is receiving water from the rainfall and from a tributary Sisarma in the west and from a few ephemeral streams and the surface runoff. Therefore, the rock weathering in the catchment area is important. Sisarma stream drains through the limestone to the lake. The lake, therefore, receives Ca and Mg by the dissolution of dolomitic limestone. The concentration of both is almost the same in the lake water and combined together are much less than HCO_3^-. Therefore, all HCO_3^- is not derived from carbonate weathering but from the silicate source rocks, which are the most abundant lithology of the drainage basin. The content of HCO_3^- in carbonate and silicate rocks is 1.518 and 2.317 meq/l respectively, indicating that more HCO_3^- is derived from silicate source than carbonates. In carbonate weathering, dissolution and precipitation kinetics of calcium carbonate are determined by the chemistry of the system H_2O-CO_2-$CaCO_3$. The chemical reaction may be given by the overall reaction:

$$H_2O + CO_2 + CaCO_3 \leftrightarrows H_2CO_3 + CaCO_3 \leftrightarrows Ca^{2+} + 2HCO_3^-$$

which involves three independent processes, CO_2 present in the aqueous solution is converted into the aggressive agent H_2CO_3. Ca^{2+}, HCO_3^-, and CO_3^{2-} are released from the mineral surface by the following reactions:

$$CaCO_3 + H- \leftrightarrows Ca^{2+} + HCO_3^-$$

$$CaCO_3 + H_2CO_3 \leftrightarrows Ca^{2+} + 2HCO_3^-$$

$$CaCO_3 + H_2O \leftrightarrows Ca^{2+} + CO_3^{2-} + H_2O$$

reaction accomplished by carbonic acid (H_2CO_3) attack (or organic acid attack, can be represented as H_2CO_3 is for calcite:

$$H_2CO_3 + CaCO_3 \rightarrow Ca^{2+} + 2HCO_3^- \text{ and for dolomite}$$

$$2\,H_2CO_3 + CaMg\,(CO_3)_2 \rightarrow Ca^{2+} + Mg^{2+} + 4HCO_3$$

Table 5.23: Chemical Characteristics of Pichola Lake

Sample No.	pH	EC ($\mu s\ cm^{-1}$)	TDS	DO	CO_3	HCO_3	Cl	SO_4	PO_4	Ca	Mg	Na	K	SiO_2
			Amount (mg/l)											
1	8.84	717	440	8.9	12	207.4	73.84	24	0.117	21.62	21.42	78.80	3.91	4.56
2	9.10	699	440	10.0	15	231.8	72.42	30	0.113	21.62	21.91	72.60	3.91	5.29
3	9.20	723	440	10.0	9	225.7	71.00	33	0.105	21.62	23.37	68.00	3.91	4.20
4	9.12	714	440	9.8	12	231.8	72.42	36	0.098	21.22	22.15	75.70	3.91	4.12
5	9.02	712	440	9.3	9	239.8	72.42	36	0.094	22.44	20.94	71.05	3.68	4.14
6	8.87	740	450	9.0	12	244.0	72.42	43	0.098	22.44	19.84	77.25	3.68	4.06
7	9.07	629	400	8.0	9	231.8	71.00	93	0.090	21.62	22.40	75.70	3.91	4.32
8	9.09	632	410	7.0	12	250.1	72.42	27	0.090	21.62	19.48	72.60	3.91	4.14
9	9.23	628	410	6.0	12	224.0	72.42	35	0.094	20.04	21.91	68.00	3.91	3.84
10	9.23	629	410	7.2	12	231.8	72.42	30	0.098	22.84	18.26	74.15	3.91	3.89
11	9.15	635	410	6.0	12	250.1	72.42	16	0.097	20.04	21.91	78.80	3.91	3.89
12	9.27	630	410	6.0	9	244.0	75.26	39	0.090	20.04	22.15	69.55	3.68	3.89
13	7.63	812	530	6.1	15	237.9	73.84	33	0.180	34.44	19.48	74.15	3.68	18.24
14	9.21	636	410	6.6	12	231.8	73.84	30	0.090	20.84	20.45	68.00	3.91	3.66
15	9.25	647	420	6.1	12	231.8	73.84	27	0.090	20.04	22.88	75.70	3.68	4.05

In soil water and groundwater, where carbonic (and organic) acid concentrations are high, the resulting concentration of Ca^{2+} and HCO_3^- are also high (Matthess *et al.*, 1992). This is observed in groundwater (well water) of Pichola Lake. In this lake there is a island over which Jagmandir temple is situated (Fig.5.40) and there is a well in the premises, the water of which show low pH of 7.63 and high $PCO_2^{10-2.18}$ compared to the partial pressure of atmospheric carbondioxide ($PCO_3^{10-3.25}$) and more of Ca^{2+} 34.44 mg/l and HCO_3^- 238 mg/l and high SiO_2 18.24 mg/l compared to lake water mentioned above (Table 5.23). The cause of pH has been studied by several workers (Garrels and Mackenzie, 1967) and can be explained fundamentally in terms of the natural processes of weathering, evaporation and CO_2 gas equilibration. In Pichola Lake the weathering has resulted in groundwater containing dissolved HCO_3^-, the concentration of which is more than twice the concentration of Ca^{2+}+ Mg^{2+}. This is because $(HCO_3)_{Si}$ is higher (2.312 meq/l) than $(HCO_3)_C$ (1.588 meq/l) indicating movement of water through silicate rocks. Therefore, upon evaporation of the ground water, after passing into a lake, the lake cannot precipitate all HCO_3^- as Ca^{2+} or Mg^{2+} carbonates, and some HCO_3^- remains behind to be concentrated by evaporation. The evaporation results in $CaCO_3$ precipitation leading to the removal of Ca^{2+} and HCO_3^- according to the reaction:

$$2HCO_3^- + Ca^{2+} \rightarrow CaCO_3 + CO_2 + H_2O$$

This is aided by loss of CO_2 resulting in the rise of pH and supersaturation and precipitation of calcite. As the inlet water continues evaporation it causes concentration of HCO_3^- to rise at about the same rate as other ions. Such as Cl^- which is not involved in precipitation reactions. During this process, pH is also rising and because of increase HCO_3, concentration reaction may take place

$$H^+ + HCO_{3-} \rightarrow H_2O + CO_2 - \uparrow$$

Because of rise in pH some HCO_3^- is converted to CO_3^{2}, the concentration of CO_3^{2-} ions also rises (Berner and Berner, 1987).

Among the cation, Na and K are dominant and contribute on average 53.15% to the total cation budget and Mg^{2+} is next in abundance contributing 28.87%, whereas Ca^{2+} is least abundant with an average contribution of 17.96% (Table 5.24), the predominance of Na+K over Mg^{2+} and Ca^{2+} is attributed to the silicate weathering as the catchment area is dominated by phyllite, orthoquartzite, and greywacke lithology. The lower content of dissolved silica is capable of influencing the dissolved silica budget in the basin, especially when environmental factors that control the growth and population of silica secreting and consuming organisms may vary. It is therefore, visualized that the source of PO_4 and dissolved silica in the well water is by the leaching of silicate and phosphate rocks, which has resulted in enrichment of both the elements in groundwater.

Table 5.24: Chemical Characteristics and percentage of equivalent per million of constituents in Pichola Lake

Sample No.	CO_3	HCO_3	Cl	SO_4	Ca	Mg	Na	K	Ca	Mg	(Na+K)	(CO_3+HCO_3)	SO_4	Cl
			Amount (mg/l)											
1	0.40	3.40	2.00	0.50	1.08	1.78	3.43	0.10	16.91	27.93	55.16	60.33	7.93	31.74
2	0.50	3.80	2.04	0.63	1.08	1.82	3.16	0.10	17.54	29.61	52.85	60.85	9.86	29.29
3	0.30	3.70	2.00	0.69	1.08	1.95	2.95	0.10	17.76	32.00	50.24	59.83	10.27	29.90
4	0.40	3.80	2.04	0.75	1.08	1.84	3.29	0.10	17.11	29.20	53.69	60.09	10.73	29.18
5	0.30	3.60	2.04	0.75	1.12	1.74	3.09	0.09	18.56	28.86	52.58	58.30	11.21	30.49
6	0.40	4.00	2.04	0.89	1.12	1.62	3.36	0.09	18.10	28.80	53.10	50.18	22.01	27.81
7	0.30	3.80	2.00	1.94	1.08	1.87	3.29	0.10	17.05	29.44	53.51	51.02	24.10	24.88
8	0.40	4.10	2.04	0.56	1.08	1.62	3.16	0.10	18.13	27.23	54.64	63.37	7.91	28.72
9	0.40	4.00	2.04	0.73	1.00	1.82	3.43	0.10	17.03	31.02	51.95	61.38	10.17	28.45
10	0.40	3.80	2.04	0.62	1.14	1.52	3.22	0.10	19.07	25.41	55.52	61.19	9.10	29.71
11	0.40	4.10	2.04	0.33	1.00	1.82	3.43	0.10	15.77	28.72	55.51	65.48	4.84	29.68
12	0.30	4.00	2.12	0.81	1.00	1.85	3.02	0.09	16.79	30.94	52.27	59.47	11.22	29.31
13	0.50	3.90	2.08	0.69	1.72	1.62	3.22	0.09	25.85	24.36	49.79	61.37	9.61	29.02
14	0.40	3.80	2.08	0.62	1.04	1.70	2.96	0.10	17.95	29.37	52.68	60.83	9.05	30.12
15	0.40	3.80	2.08	0.56	1.00	1.91	3.29	0.09	15.92	30.29	53.79	61.39	8.22	30.39

Udaisagar Lake

The Udaisagar Lake situated towards the east of Udipur city, covers an area of about 10.36 km^2, is 4.02 km in length, 3.22 km in width and is 10.6 m in depth. It is mainly fed by the Ahar River flowing from the west and meeting the lake towards its western margin. The surplus water of the lake is drained off through Berach river in the ENE (Fig.5.41). The catchment drainage basin of this, and other Rajasthan lakes are small with small streams draining into the lake and there is evaporational loss of water due to high temperature particularly in hot summer.

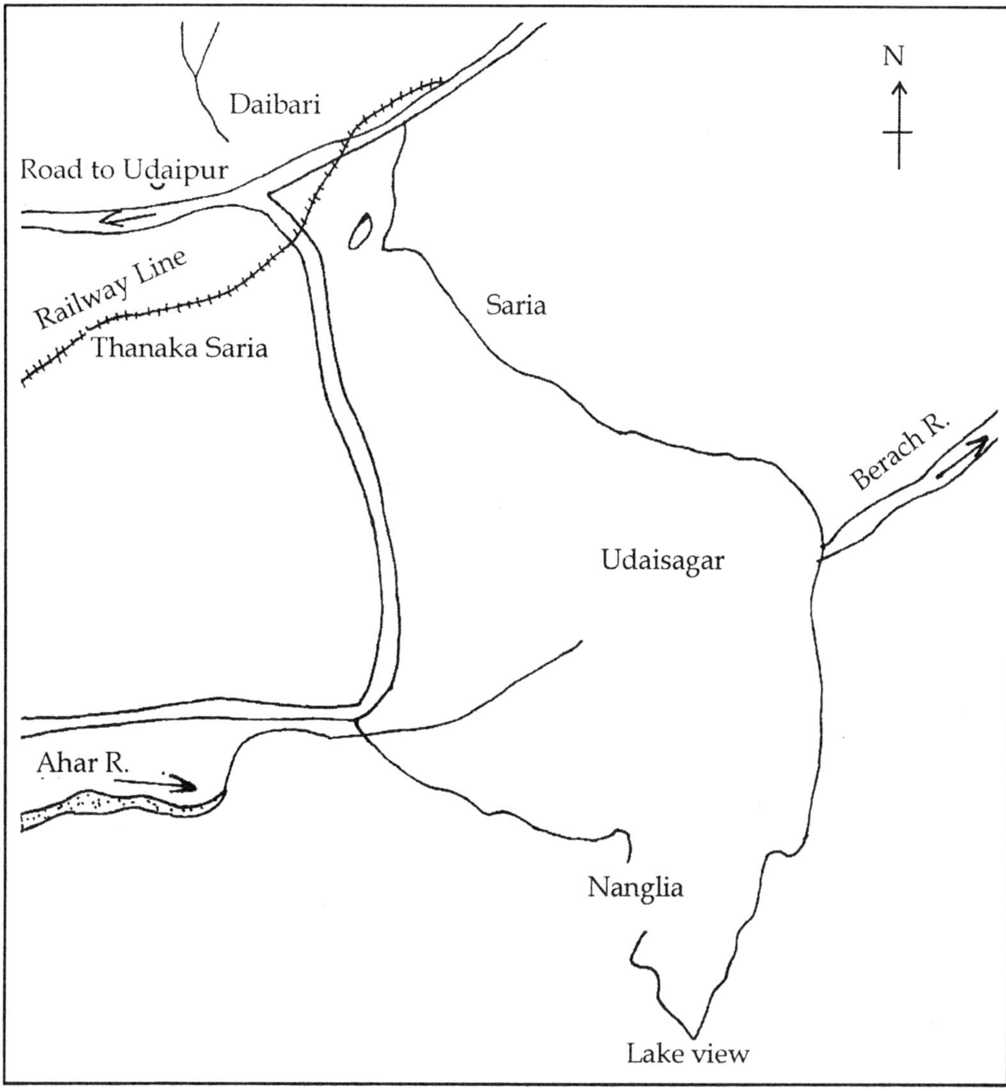

Fig. 5.41. An outline map of Udisagar Lake showing Ahar River flowing from the east and feeding the lake and Berach River in the northeast which drains out the water of the lake.

The lithology around Udaisagar Lake is carbonaceous phyllite, dolomitic limestone, and post Aravalli granites (Heron, 1953; Roy, 1988) (Fig.5.39). The phosphate deposits occur in the carbonate sequence near Matoon, Kanpur and Karbaria-Ka-Gura near the lake and a large phosphorite deposit of Jhamar Kotra occurs towards south of the lake. In case of former, dolomitic marble, calcareous grit, quartzite, phyllite and phosphorite of Aravalli Group are the main rock types. The marble is stromatolitic with which phosphorite is associated. An average annual rainfall of 635 mm has been reported by Rajasthan Pollution Prevention and Control Board (1986).

The water chemistry of Udaisagar Lake shows variation in pH of the lake at different sampling points. In general, the water at a distance from inlet river Ahar is alkaline and pH ranges from 9.26 to 9.55 whereas those which are close to inlet river are very mildly alkaline to neutral with pH ranging from 7.02 to 7.40 (Table 5.25). The E.C is correspondingly less in samples with high pH ranging from 1207 µs/cm to 1315 µs/cm while in low pH samples, it varies from 1297 µs/cm to 1355 µs/cm (Das, 1999). This shows that the low pH of waters near the inlet river is due to enhanced respiration, organic decay caused by the input of untreated pollutants brought by Ahar river and added to the lake. This resulted in the elevation of CO_2 in the lake water increasing PCO_2 and hence decrease in pH as compared to the water in other parts of the lake at a distance from the inlet. The higher partial pressure of carbondioxide log $PCO_2^{-1.59-1.87}$ compared to the atmosphere log $PCO_2^{-3.5}$ reaffirm the above phenomena and is supported by the higher E.C in the water samples, as more of Ca and other ions will be dissolved in the water. Such waters are undersaturated in Ca^{2+} and Mg^{2+} and these are not precipitated. The PCO_2 in other samples are lower than the ambient carbondioxide pressure and is reflected in the chemistry of the water. The water is supersaturated in both Ca^{2+} and Mg^{2+} in these samples.

The Ca^{2+} + Mg^{2+}, on an average contribute 43.46% to the total cation budget and Mg^{2+} alone contribute 32.33% which is almost two times more than Ca^{2+} 19.29%. Among the anions (CO_3^{2-} + HCO_3^-) contribute 40.01%, which is almost the same as Ca^{2+} + Mg^{2+} combined together (Table 5.26). HCO_3 is derived from both carbonate and silicate minerals and the high content of bicarbonate (more than twice of Ca^{2+} + Mg^{2+}) may indicate its source from silicate minerals. Because the Ca^{2+} and Mg^{2+} derived from carbonate minerals should be equal to carbonate derived HCO_3^-, according to the stoichiometric requirement of carbonate weathering reaction given above and this condition is not fulfilled. The predominance of silicate rocks, in the catchment area and computation of $(HCO_3)c$ and $(HCO_3)_{Si}$ gives average value of 1.654 and 2.074 meq/l respectively, which is in conformity with the weathering of silicate rocks in the watershed and hence does not indicate all the bicarbonate been derived from carbonate rocks. The bicarbonate alkalinity in this lake is 3.86 meq/l as compared to Pichola Lake 4.20 meq/l and Fatehsagar Lake 4.58 meq/l on an average.

Table 5.25: Water chemistry of Udaisagar Lake, Rajasthan

Sample No.	pH	EC (μscm^{-1})	TDS	DO	CO_3	HCO_3	Cl	SO_4	PO_4	Ca	Mg	Na	K	SiO_2
						Amount (mg/l)								
US-1	9.26	1270	700	3.4	0.00	231.8	144.84	93	0.178	20.84	34.09	109.70	14.43	7.64
US-2	9.31	1307	710	4.8	6.00	231.8	144.84	86	0.190	19.23	34.57	113.90	14.43	3.52
US-3	9.55	1299	700	3.0	0.00	244.0	143.42	89	0.178	16.83	37.01	105.10	14.43	6.92
US-4	9.54	1315	710	3.8	6.00	207.4	143.42	83	0.205	17.63	35.06	105.10	14.43	5.26
US-5	8.96	1313	720	4.1	0.00	219.6	144.84	83	0190	20.04	34.57	106.10	14.43	5.40
US-6	7.18	1331	720	4.3	0.00	244.0	143.42	110	0.190	16.83	34.09	113.90	14.43	5.48
US-7	7.40	1355	730	5.2	6.00	225.7	143.84	93	0.183	18.43	36.52	113.90	14.43	7.80
US-8	8.86	1343	730	6.8	6.00	231.8	144.84	83	0.150	18.43	35.55	111.25	14.43	5.86
US-9	9.24	1313	730	3.0	0.00	225.7	143.84	53	0.194	19.23	35.55	113.90	14.43	5.27
US-10	7.30	1332	720	10.1	0.00	231.8	143.42	45	0.135	20.04	34.57	113.90	14.43	5.38
US-11	7.12	1318	710	5.0	0.00	237.9	144.42	83	0.150	20.84	33.60	116.55	14.61	5.34
US-12	7.02	1332	720	4.3	0.00	231.8	144.84	79	0.190	18.43	34.57	126.15	14.43	7.60
US-13	7.25	1297	700	2.2	6.00	231.8	146.26	83	0.255	24.03	33.11	116.55	14.43	6.42

Table 5.26: Chemical characteristic and percentage of equivalent per million of constituents in Udaisagar Lake.

	Amount (meq/l)								Cations (%)			Anion (%)		
S.No.	CO_3	HCO_3	Cl	SO_4	Ca	Mg	Na	K	Ca^{2+}	Mg^{2+}	(Na^++K^+)	$(CO_3^{2-}+HCO_3^-)$	SO_4^{2-}	Cl^-
US-1	-	3.89	4.08	1.93	1.04	2.84	4.76	0.37	11.55	31.48	56.97	39.32	19.53	41.15
US-2	0.20	3.80	4.08	1.79	0.96	2.88	4.95	0.37	10.49	31.43	58.08	41.39	17.88	40.73
US-3	-	4.00	4.04	1.85	0.84	3.08	4.56	0.37	9.84	34.79	55.73	40.43	18.74	40.83
US-4	0.20	3.40	4.04	1.72	0.88	2.92	4.56	0.37	10.08	33.42	56.50	38.43	18.45	43.12
US-5	-	3.60	4.08	1.72	1.00	2.88	4.63	0.37	11.27	32.41	56.32	38.61	18.27	43.12
US-6	-	4.00	4.04	2.29	0.84	2.84	4.95	0.37	9.54	31.55	59.11	38.73	22.17	39.10
US-7	0.20	3.70	4.04	1.93	0.92	3.04	4.95	0.37	10.08	32.77	57.15	39.49	19.61	40.90
US-8	0.20	3.80	4.08	1.72	0.92	3.04	4.83	0.37	10.04	33.20	56.76	40.79	17.62	41.59
US-9	-	3.70	4.04	1.10	0.96	2.96	4.95	0.37	10.49	32.32	57.19	39.05	13.08	47.87
US-10	-	3.80	4.04	0.93	1.00	2.88	4.95	0.37	10.88	31.29	57.83	43.30	10.67	46.03
US-11	-	3.80	4.08	1.64	0.94	2.88	5.48	0.37	9.77	29.75	60.48	39.90	17.27	42.83
US-12	0.20	3.80	4.12	1.72	1.20	2.75	3.39	0.37	19.55	35.73	48.72	40.62	17.55	41.83

The (Na^+ + K^+) together contribute 56.88% on an average to the cation budget of this lake (Table 5.26). The high Na^+ and Cl^- (42.39% av.) are due to aridity of climate involving high evaporation, poor rainfall and limited inflow of water in the basin-a characteristic of desert region and anthropogenic pollution.

The high Cl^- content is primarily due to alkaline/saline soils and saline groundwater in the drainage basin and direct sewage discharge in the lake. According to Meybeck (1979), 30% of Cl^- in river water arises from pollution. Domestic sewage contains considerable Cl^- and if it is discharged in lake or river it adds a good percentage of it. Similarly, about 28% of Na^+ is added to river water through pollution.

The Plot of TDS (mg/l) against Na/(Na+Ca) (wt. ratio) shows that water chemistry fall in the middle but outside the boomerang (Gibbs, 1970) because the value of Na/(Na+Ca) is high (Fig 5.42). According to Gibbs, intermediate TDS and low Na/(Na+Ca) is the location of water whose composition is controlled by rock weathering. The high Na/(Na+Ca) values of Udaisagar Lake has led them to shift outside the boomerang though it may indicate that the lake composition is controlled by rock weathering and high Na/(Na+Ca) is due to evaporation in this arid area. Meybeck (1984) has mentioned that total dissolve solids (TDS) in rivers draining sedimentary rocks (including carbonates) tends to be atleast twice the TDS in rivers draining only crystalline (igneous and metamorphic) rocks (Holland, 1978).

The high sulphate content 17.58% of the total anion budget is derived from black carbonaceous phyllite which contains pyrite and weathers readily to H_2SO_4, which then reacts with silicate and carbonate minerals in the enclosing rock to release cations (particularly Ca^{2+}) and SO_4^{2-} into river water. The significant sources of sulphate, are rock weathering (33%), pollution (43%), and natural biogenically derived sulphate in rain (17%) (Berner and Berner, 1987).

The phosphate content in this lake is 186 μg/l, whereas in other lakes viz., Fatehsagar and Pichola, it is 91.25 and 102.93 μg/l. The concentration of phosphate in lakes and rivers is generally low (60 μg/l). The primary sources of phosphorus in lakes are direct rainfall and snowfall on the lake and runoff from the surrounding drainage area. In oligotrophic lakes most phosphorus in runoff comes from rock weathering and soil transport. However, in areas influenced by humans there are additional sources of phosphorous including agriculture, runoff (containing phosphorous from fertilizers and animal wastes) and sewage (containing phosphorous from human wastes, detergents, and industrial wastes), which are discharged directly into the lake or its inlet tributaries. In Udaisagar Lake, the main source of phosphate is from rock weathering and mining of phosphorite. The dolomitic limestone contains seams of phosphorite, and mine tallings heaped near the querry are traversed by

Ahar River during its journey, which carries phosphate to the lake. Additionally, it is also the carrier of human waste, which drops in the Udaisagar Lake. Atmospheric precipitation is not significant due to deficiency of rainfall and aridity of climate. The low dissolved oxygen (DO) at 4.15 mg/l indicates increased organic content due to high nutrient load such as phosphate and sulphate in the lake and development of anoxic conditions leading to eutrophic conditions in the Udaisagar Lake. The lake therefore, poses a great environmental threat to the local population and aquatic life.

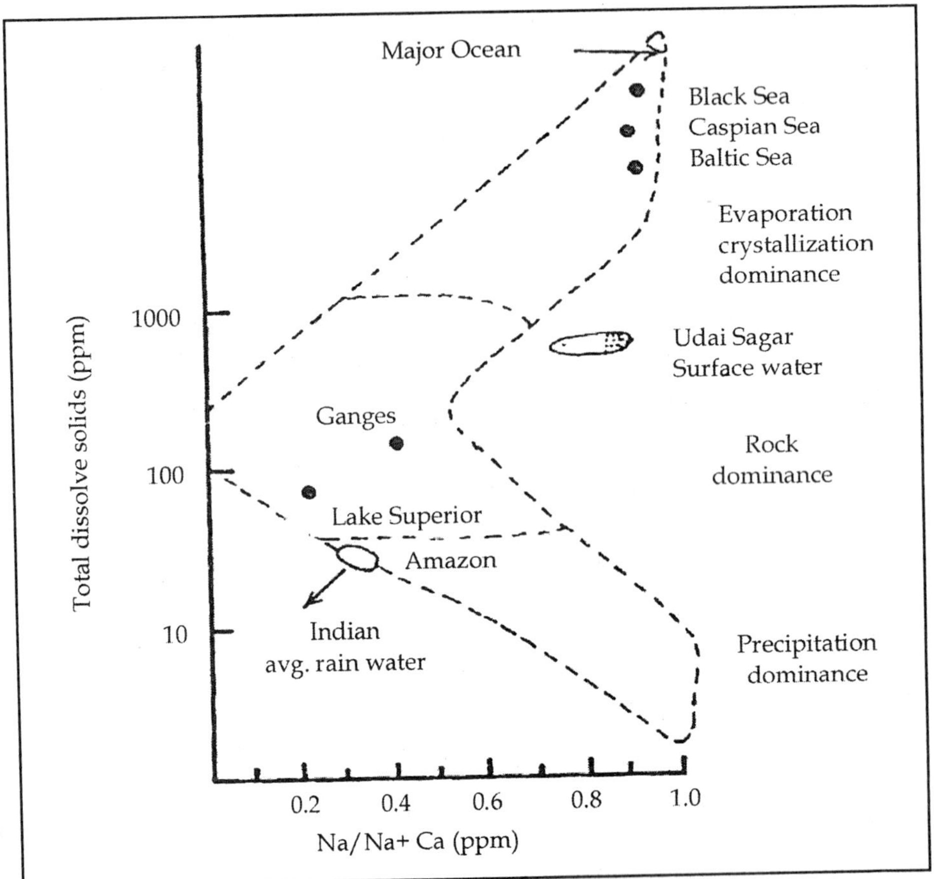

Fig. 5.42. In a Gibb's (1970) diagram Udisagar Lake water plots outside the rock dominant compartment but in the same area. The high Na in water has shifted plots outside the domain of rock dominance.

Salt Water Lakes

Tsokar Lake of Higher Himalaya

Most of the high altitude lakes of Himalaya are saline due to low precipitation and high evaporation. The higher Himalayan regions are rain shadow areas and cold desert conditions prevail resulting in high evaporation during the day time.

Tsokar Lake is about 160 km from Leh in district Ladakh, Jammu and Kashmir, at an elevation of 4535 m above msl. It is geographically bound between latitudes 33°15′ 60″ to 33° 22′ 00″ N and longitudes 77°55′ 00″ to 78° 05′ 00″ E and two mighty Himalayan ranges Zanskar in the south and Ladakh in the north rises in this region. The area where Tsokar is situated is facing a very cold and arid arctic type of climate with temperatures varying from sub-zero during the night to 28°C during the day. The Tsokar is at present very shallow with a maximum depth of about 12.5 m and is assumed to be a remnant of the original lake with an area of 136 sq km (Rastogi, 1979). To its south is another lake, the Startspuk Tso which is a fresh water lake and feeds Tsokar through a meandering channel (Fig.5.43). The lake is fed from the nullahs in north and south of the lake, the northern nullahs are mostly dry but show well developed braided stream channel features while southern nullahs are almost perennial.The southern nullahs, Polokong Ka Phu and Nuruchan Lungpa meet Startspuk Tso directly, from where the water is drained into the Tsokar through a meandering channel about 6 to 8 m in width and about 2.5 km in length. In addition to these, there are a number of springs within the lake basin which contribute directly to the Tsokar Lake. However, it is estimated that the total contribution from these springs is much lower as compared to the total recharge of water into the lake. The rate of evaporation is extremely high.

Fig. 5.43. Lanset imagery showing Tsokar Lake and its gradual desiccation. In the south fresh water lake Startspuk flows and feeds Tsokar Lake.

The rocks exposed around the basin include the members of the Puga Formation (Precambrian), Samdo Formation (Miocene). To the north-west of Tsokar, limestones, shales, and sandstones are the predominant lithology. The ranges north and south of the lake comprises mostly phyllite and slate with minor limestone bands of Precambrian age and to the east of the lake mostly biotite, tourmaline and porphyritic granite (Polokongkala granite) of upper Palaeozoic age are exposed.

The bank of the lake mostly comprises clay, silt and mud with fragments of rocks derived from adjoining hills and adjacent outcrops. From the salt encrustations near the shore, circular crater shaped structures with dark sludge at the base are reported and the water close to these structures is luke warm, these therefore, indicate a point of emergence of hot springs (Rastogi, 1979). The area is located very close to Indus suture where the Indian plate meets the Tibetan plate and along the suture zone the ophiolites (Samdo Formation) and the molasse facies of the Indus Formation are exposed. The basin is bounded by two longitudinal faults and forms a graben structure where the central block has subsided to constitute the basin.

The chemistry of water in the lake varies along the length and breadth of the lake due to concentration of brine. Since the terrain is difficult, the previous investigations are restricted to Geological Survey of India, considering the lake to be of economic potential. However, the geochemical data of brine generated by them shows large variability and is not comparable to the measurements made presently (Table 5.27), because of variation in location of sample, season of collection and change in morphology of the lake. The pH of the lake water, from different sampling points, show variation from 7.56 to 8.74 and E.C also show large variation from 340 to 8350 μs/cm. This is due to dissolved ions concentration in water which varies from place to place. The CO_3^{2-} and HCO_3^- concentration are high and also highly variable, particularly, the very high carbonate concentration is due to salt in the lake. In natural fresh waters, the CO_3^{2-} concentration is small compared to the HCO_3^- concentration. The relatively low concentration of Ca^{2+}, Mg^{2+} and CO_3^{2-} and HCO_3^- compared to SO_4^{2-}, Cl^-, Na^+ and K^+ suggest that water is not rich in carbonates (Table 5.27). The SO_4^{2-} concentration is 2.4 times higher than the Cl^-. In most of the samples, even higher concentration is recorded. The concentration of Ca^{2+} and Mg^{2+} is also higher but lower than Na^+ and K^+. The Mg^{2+} is ~ 3 to 4 times higher than Ca^{2+} and Na^+ is 2.67 times higher than K^+. The Na: Cl ratio is 2 : 1, which may lead to the formation of halite in the brine. The alkalinity of water is generally high and is more than Ca (alk > Ca). Though the chemistry of water is highly variable, but a very broad generalization of solutes in water can be decerned i.e. Na > SO_4 > Cl > Mg > K > Ca > CO_3 > HCO_3. The major brine types reported from different parts of the world are (a) Na- CO_2-Cl-SO_4, (b) Na-Cl-SO_4, (c) Na-Mg-Cl-SO_4 and (d)

Table 5.27: Chemical characteristics of Tsokar Lake, Leh district (J & K).

Sample No	pH	EC (µ scm^{-1})	CO_3	HCO_3	Cl	SO_4	NO_3	F	PO_4	Ca	Mg	Na	K	SiO_2	TDS
	Amounts mg/l														
1	7.70	827	-	439	43	-	0.87	1.16	-	49	39	60	25	-	579
2	8.47	8350	54	360	1329	1670	5.59	1.40	ND	51	206	1140	440	21	5845
3	7.56	1157	-	671	50	-	1.32	1.40	-	27	1.06	48	23	-	8099
4	8.65	340	14	152	70	10	1.96	ND	ND	38	16	4.2	1.8	9.0	238
5	8.60	1863	1320	1525	2483	59950	41	1.40	0.34	1680	5680	31000	11500	1.4	1304
6	8.74	1882	1800	2135	23362	63400	37	1.08	3.96	1680	6189	31000	11500	1.3	13174
7	8.70	1863	1825	582	24830	59990	51	0.75	3.80	1700	6700	31000	11500	1.7	1304
8	8.73	1881	1525	1247	26268	65560	38	0.75	3.72	1700	6700	29000	11000	0.9	1317
9	8.72	1884	2070	28	26300	65000	42	0.55	4.27	850	6700	3000	11000	605	1319
10	8.12	605	-	235	51	9.0	21	0.75	103	42	18	47	9.0	14	4235
11	7.4	200	-	83	7.3	39	2.4	ND	0.19	34	3.0	3.4	2.3	15	140

ND =Not determined
(unpublished data taken from Dr. S.C. Dhiman)

Ca-Mg-Na-Cl brines (Eugster and Hardie, 1978). However, notable deviations occur in many evaporite deposits. This also depends on the rock types outcropping around the drainage area.

The investigation carried out by many scientists and institutions have not reported evaporite units in the rock formations of drainage basin, therefore, the composition of the brines will suggest its source of supply. The previous and present investigations of Tsokar Lake have revealed that the present lake has resulted through evaporation and concentration of lake waters of once a much larger lake. The estimated composition of Tsokar brine indicates Na-SO_4-Cl type evaporative course, similar to the Poison Lake, Washington (Hardie and Eugster, 1970).However, it differs only in one factor, that is in the potash enrichment, which together with sodium forms a major constituent of the Tsokar brine. When the waters get concentrated, Ca values drop down and also the CO_3 values decrease along with precipitation of $CaCO_3$. Consequently, the brine gets enriched in sulphate and chloride. Carbonate precipitation may also occur at higher salinities upto around 80 percent (Schreiber, 1998). The relative abundance of Na and Cl is much higher in comparison to Ca and Mg thereby resulting in the precipitation of halite at much lower salinities. However, X-ray analyses of the precipitated salts and the bottom sediments of Tsokar did not confirm gypsum as mineral phase (Rastogi, 1979). It is surmised that due to very low concentration of calcium after its precipitation as $CaCO_3$, it is depleted in brine, thus leading to the precipitation of sodium sulphate as is evident from high Na^+ and SO_4^{2-} in the brine.

Sambhar Lake of Rajasthan

Sambhar is the largest inland saline lake of India covering an area of approximately 225 sq km in central Rajasthan (Fig.5.44). Located about 80 km north-west of Jaipur in north western India, it is the most important, extensive and the largest playa of the Thar desert, Rajasthan. The lake basin is bounded to the north-west and the west by the ~ 500 m high Aravalli hills of Early and Middle Proterozoic age, which also divide the region into a subhumid eastern part and semi-arid western part (Thar desert). The Sambhar Lake falls in the rain shadow from south west monsoon with an annual rainfall of 100-500mm. The average annual temperature is 23°C with a maximum of 45°C. The basement below the lake floor is made up of Aravalli Schists. The maximum length of the Sambhar Lake basin is 22.5 km, while the width ranges from 3.2 km to 11.2 km, with its main axis along NE-SW directions. It is a shallow basin reaching only about 3 m at its deepest, with an average depth of 0.61m and it dries out completely in summer months. The total catchment area of the lake is 7560 km^2, most of which lies to the north and north east. Two major streams namely, Mendha and Rupangarh feed the lake in addition to numerous rivulets and surface runoff. In spite of a large catchment area, the Sambhar Lake presently receives very little runoff, partly due to present climatic conditions and partly due to human interference, i.e. small check dams along the rivers and diversion of lake water to reservoirs for salt production.

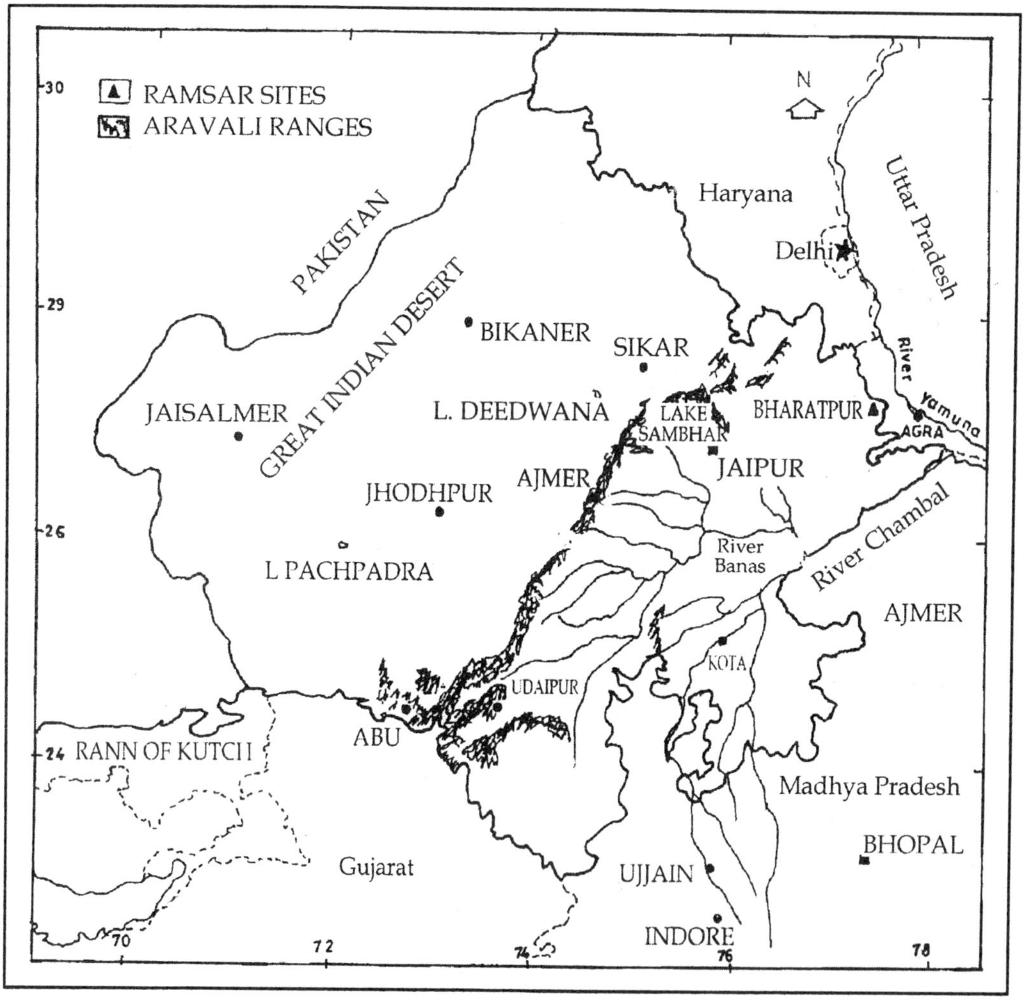

Fig. 5.44. An outline map of Udaipur showing location of Sambhar Lake NW of Jaipur, Aravali Range is also shown in the map.

The surface of Sambhar Lake presently undergoes complete desiccation every summer forming an efflourescent crust (Sinha and Raymahashay, 2004). This crust essentially consists of halite and calcite with minor amount of dolomite, carnallite, polyhalite, and sylvite (Roy *et al.*, 2001). It dissolves when it comes in contact with fresh runoff during the next rain and this process increases the solute load of the lake brine.

Earlier views suggested a wind-borne source of salt from the Rann of Kutch (Holland and Christie, 1909), remnent of inland sea (Tethys) during the Tertiary (Godbole, 1952), and dissolution of halite bed in the lake area (Bhattacharya *et al.*, 1982). However, recent isotopic investigation of lake water has discarded marine origin and has suggested that the lake brine is completely replenished by precipitation and surface runoff. Considering this as the source of alkalinity, most of the work has discussed the composition of lake brine and local water bodies. Recently, it is realised that the complete picture of the evolution of this lake will emerge on the basis of study of the nature of rock weathering in the catchment area together with a detailed mineralogical data of evaporite (Sinha and Raymahashay, 2000, 2004).

Weathering of primary minerals like feldspar, biotite and amphibole to clay and iron oxyhydroxides, releases K^+, Na^+, Ca^{2+}, Mg^{2+} along with bicarbonate and silica to river and groundwater. The CO_2-saturated rainwater is primary weathering agent. The major ions and silica are released into river and/or groundwater by an exchange reaction with H^+ ion derived from dissolved CO_2 (H_2CO_3) and Al is locked up in the clay mineral and ferrous ion (Fe^{2+}) after being oxidized to Fe^{3+} forms goethite. These primary and secondary minerals in the catchment area of two main inlet rivers, of Mendha and Rupangarh, are the source of these dissolved constituents in the Sambhar Lake.

During the progressive evaporation of water in a saline lake, the sequence of minerals precipitated follows the chemical divides proposed by Eugster and Hardie (1978). During the early stage of evaporation increasing salinity precipitates calcite and high Mg-calcite. Subsequently, the relative concentration of Ca, Mg, HCO_3, SO_4 and Cl control the sequence sulphates, silicates and chlorides (Fig. 5.45) (Sinha and Raymahashay, 2004). Carbonates are the most wide spread evaporites in the sediments occurring at all depths but are particularly dominant in the lower most units. Calcite is the main carbonate mineral and some horizons have dolomite in significant quantities. The work of Yadav (1995) showed that in the present day surface and subsurface water of the Sambhar Lake area, the ratio mCa/mMg is less than $Kc^2/Kd = 2:1$ where Kc and Kd are solubility products of calcite and dolomite, respectively. This will favour dolomitisation of calcite according to the reaction;

$2CaCO_3 + Mg = CaMg(CO_3)_2 + Ca$. A strong correlation between dolomite content, MgO/CaO ratio in insoluble fraction and $\delta^{18}O$ of carbonate fraction in the Sambhar Lake sediments (Sinha and Smykatz-Klosss, 2003) lend support to the above idea.

In the core sediment below a depth of 5.5 m, an evaporite assemblage with gypsum, calcite, dolomite, halite and minor thenardite is encountered, whereas in sediments in the upper 5.5 m gypsum is conspicuously absent and

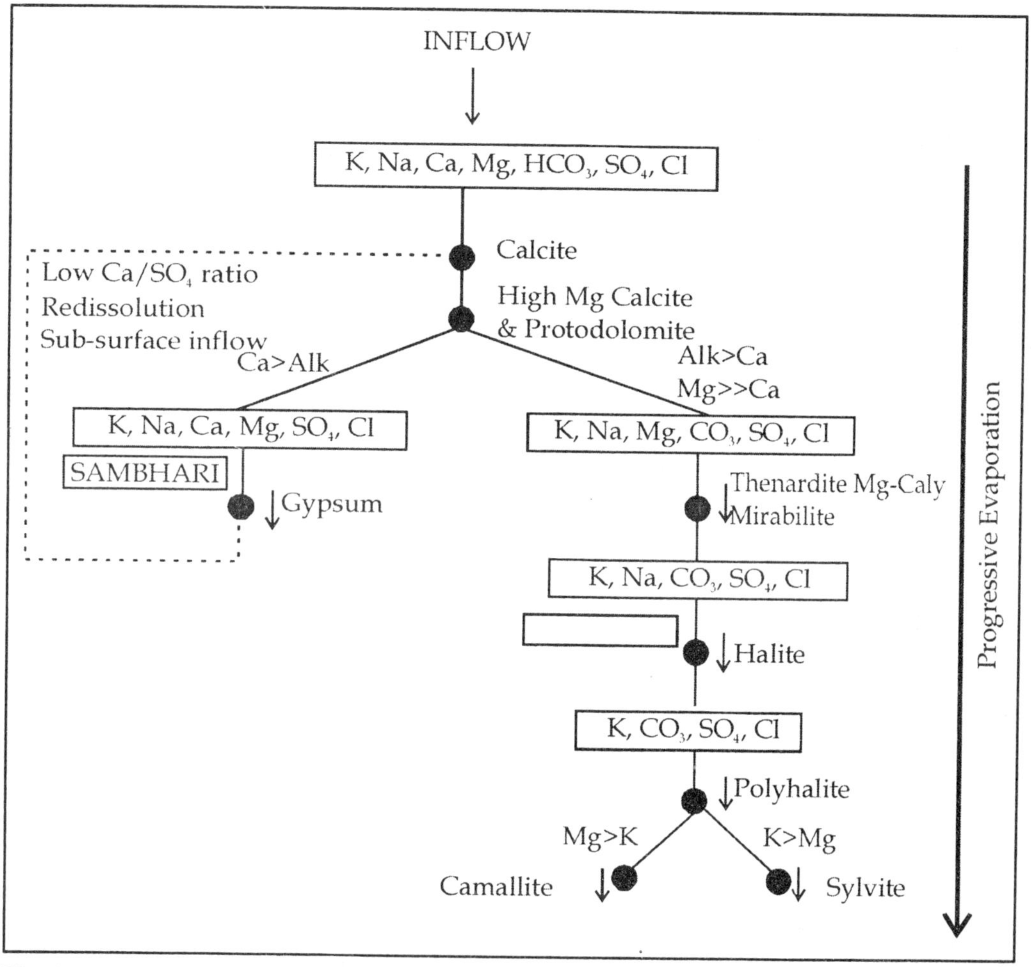

Fig. 5.45. Diagramatic representation of evolution of brine (after Sinha and Raymahashay, 2004)

the sulphate mineral assemblage is represented by thenardite with or without kieserite. This suggests a change in the brine evaporation chemistry corresponding to this depth. This indicates that during an early stage of lake evolution when deeper sediments were deposited, Ca concentration in the brine was higher than alkalinity on an equivalent basis. The brine evolution therefore, followed a left branch of the chemical divide shown in Fig 5.45 depositing calcite and gypsum. As gypsum continued to precipitate, the principle of binary salt formation (Drever, 1997) would dictate that SO_4 progressively increased while Ca decreased to a low value. At this stage, the evaporation path looped back to carbonate precipitation and followed the right branch of the chemical divide in Fig 5.45. Further evaporation under the condition of alkalinity greater than Ca and Mg led to the precipitation of Mg-clay, i.e. chlorite-smectite mixed layer together with thinardite in the younger sediments (Sinha and Raymahashay, 2004). At the Sambhar Lake, the evaporite-rich horizons possibly developed under conditions when evaporation was high and groundwater inflow was sufficient as suggested by Rosen (1991,1994). A pronounced variation in the evaporite mineral assemblage in the stratigraphic column represented by the core points to sharp temporal changes in the chemical environment of the lake. Such chemical transformations, with time have been frequently observed in saline lakes around the world. Several reasons have been suggested for such transformations, including change in brine chemistry related to source area and/or climate, chemical stratification of lakes, and recycling of salts by fractional dissolution of efflourescent crusts. In Sambhar Lake, such changes could be climatically driven. The total time span involved in the formation of these deposits is more than 30 Ka, and therefore Quaternary climatic fluctuations may be the prime factor in pronounced salinity variations in the Sambhar Lake.

Lonar Crater Lake

Lonar Crater Lake, in the Buldana District of Maharashtra, India (19° 58' N, 76° 31' E), is an almost circular depression in the Deccan Traps. It has an average diameter of 1830 m and a depth of about 150 m, with a saline lake occupying the floor of the crater (Nandy and Deo, 1961; Chowdhury and Handa, 1978; Fredriksson *et al.*1973; Nayak, 1972). The rim of the crater is raised 20 m above the surrounding plain. Lonar is the only well-known example of an impact into a (continental) flood basalt terrain. The origin of this crater has been a matter of controversy since the 19th century. Most early investigations ascribed it to some form of volcanic explosion or subsidence. The first suggestion that Lonar Crater was formed by meteoritic impact seems to have been made by Lafond and Dietz in 1964. Now there are abundant evidence in support of an impact origin of this crater, including the relative youthfulness of the crater (in comparison to the age of regional volcanism), its morphology, the presence of glasses around the crater, ejected melt breccias with shocked minerals, and a subsurface breccia identified-in drill cores beneath the crater (Nayak,1972; Fredriksson *et al.*, 1973, 1979; Fudali *et al.*, 1980).

About 700 m north of the rim is a shallow depression outlined by a raised rim 300 m in diameter; is assumed to be second crater formed together with Lonar Crater, perhaps by a fragment of the same impactor (Fredrikson *et al.*, 1973).

The circular outline of the rim is broken by a prominent creek in the north eastern part where a perennial spring known as Dhara with appreciable discharge, issues near the top of NE side of the crater. The walls are quite steep and in most places slopes at angle varying from 25° to 35°. The lake is closed one without any outlet, and is fed by streams, the sources of which, with exceptions, are confined practically within the rim of the basin. The depth of water in the lake varies considerably from place to place as well as with the season, the maximum depth of 5.5 m being recorded during the monsoon. In the pre-monsoon summer months the depth of water is much lower, the minimum depth recorded as 0.15 m in April, 1953. The drillings in the lake bed have shown that it is horizontal with about 30 m of silt resting on about 50 m of brown to grey highly weathered rock.

The chemical composition of lake water collected from different parts of the lake is shown as an average in Table 5.28 as no variation is found in them. In this table, water analysis of samples from the three springs, which flow, into the lake is also shown.

The water analysis data (Table 5.28) shows that the lake water is alkaline, having pH around 9.7-9.8. The sodium ion constitutes over 95% of the total cations (in meq/l) in the lake water, whereas amongst the anions, chloride, carbonate and bicarbonate ions are dominating. The Sulphate and nitrate contents are much lower. The Lonar Lake water is therefore Na-Cl-CO_3-HOC_3 type.

A comparison of the ratios (by weight) of the various constituents present in the lake water with that of the springs, which discharges their water into the lake, gives an idea of the changes taking place in the chemical composition of lake water with the passage of time (Chawdhury and Handa, 1978). There has been large enrichment of Na relative to Ca and Mg. There is also an enrichment of Na and impoverishment of K in the lake water as compared to the influent spring waters. Reduction in K contents is due to its adsorption in the clay minerals. Further, a major portion of Ca and Mg ions supplied by these spring waters have been removed from the lake waters, through precipitation as their carbonates. The nitrate in the lake water is derived from the spring influents into the lake. However, the NO_3 contents of the influent spring waters and that of the lake waters are not constant but nitrate content of the lake water is generally much lower than that of the spring water (Table 5.28).

Presumably bacterial reduction of NO_3- ions to ammonia is taking place, a fact which is also supported by the presence of around 1.9 ppm of free ammonia in the lake water (Table 5.28). The role played by the ammonia in the geochemical reaction in the lake water is very important. The high pH caused by the formation of ammonia has apparently directly affected the CO_3-HCO_3 equilibria and precipitation of calcium and magnesium took place.

Table 5.28: Chemical composition of water of Lonar Lake and Influent spring waters (in mg/l) and E.C in μs/cm at 25°c.

Parameters	Lonar Lake		Dhara		Influent Springs Sitanahani		Ramgaya	
	Aug.70	June 71	Aug 70	June71	Aug.70	June71	Aug70	June71
pH	9.70	9.85	7.80	7.20	7.75	7.30	8.05	7.50
E.C.	20900	20500	780	1550	560	525	790	740
TDS	14440	16170	568	1200	370	382	548	543
CO_3	2006	2547	-	-	-	-	-	-
HCO_3	2158	1876	315	345	370	358	390	365
Cl	4936	5505	47	100	10	12	39	41
SO_4	124	129	19	67	4.3	4.8	14	14
NO_3	22	6.1	96	409	11	13	73	82
Ca	7.8	8.9	58	125	32	33	41	45
Mg	12	13	44	97	36	36	43	42
Na	5579	6252	40	92	46	47	72	75
K	23	15	1.6	3.6	0.8	0.7	1.6	1.7
SiO_2	20	24	56	57	54	59	53	5.9
Free ammonia	-	1.9	-	-	-	-	-	-

(Taken from Chawdhury and Handa, 1978)

Table 5.29: Chemistry of Lonar Lake sediments and Deccan Trap

Wt.% oxides	Lake Sediments				Trap rock	
	M1 3.45m	M3 3.55m	M4 2.45m	M9 0.40m	Deccan Trap (Sengupta unpub.)	Weathered* bottom siltbed
SiO_2	41.54	40.16	47.76	42.02	49.71	42.49
Fe_2O_3	10.97	11.16	12.35	11.98	3.84	11.35
FeO	-	-	-	3.18	9.02	-
Al_2O_3	11.19	10.58	11.24	10.57	12.23	16.85
TiO_2	1.56	1.44	2.69	2.25	2.56	-
P_2O_5	0.40	0.46	0.32	0.27	0.16	-
MgO	7.44	5.38	7.43	7.05	5.55	4.70
CaO	10.50	7.33	13.55	11.45	10.36	9.60
Na_2O	3.64	4.10	2.49	1.62	1.76	tr
K_2O	0.75	0.64	0.18	0.24	0.54	tr
MnO	0.16	0.16	0.22	0.21	0.19	-
Cl*	0.38	0.70	0.19	0.16	0.00	0.03
SO_4*	0.06	0.12	0.05	0.00	-	0.33
CO_2	2.64	1.06	2.44	1.32	0.39	-
$H_2O(-)$	6.77	6.24	2.54	4.74	0.90	-
$H_2O(+)$	2.25	6.24	1.90	3.08	2.26	-
Org.Matt.	0.69	3.10	2.84	2.70	-	-
LOI	-	-	-	-	-	13.40

*Water soluble (Taken from Chawdhury and Handa, 1978).

Some changes in the lake water in respect of increase in CO_3 and decrease of HCO_3 and decomposition of NO_3 are related to evaporation. However, very little seasonal variations in the chemical composition of two spring water is noted, except that spring "Dhara" which has much higher discharge than the other two, showed more salt content in the dry season.

The composition of the sediments from the bottom of the lake does not show variation and is very similar to that of the Deccan Trap (Table 5.29). Slight weathering of the trap rocks in the crater wall may lead to composition of the lake sediments as estimated by analysis. The lake sediment contain very small amounts of water i.e. soluble chloride and sulphate. This rules out the possibility of any source of salt from the bottom.

The salinity of the Lonar Lake water is due to accumulation over the years through the process of evaporation in the closed basin. The inflowing springs and in situ rock weathering within the crater area have controlled the composition of salts. According to Nandy and Deo (1961), the alkalinity of the lake is due to the meteoric waters feeding the lake. The influents, while passing through volcanic rocks, which are source of sodium salt, dissolve the soluble salts and bring them to the lake. Evaporation and interactions of the salts in the lake water resulted in the concentration of salts.

Chapter 6

SEDIMENT CHEMISTRY GENERAL

Sedimentation precisely involves an interaction of the atmosphere and hydrosphere on the crust of the earth. Various aspects of sedimentation are related to weathering, erosion, deposition, and diagenesis, but none of the processes work in isolation. The original constituents of the crust, the igneous rocks are mostly unstable with respect to the atmosphere and hydrosphere. They have been formed at a high temperature and pressure, and therefore, they do not remain stable under the very different conditions at the earth's surface. Among the common minerals of igneous rocks, only quartz is highly resistant to weathering processes. All the other minerals are susceptible to alteration by action of oxygen, carbonic acid and water, to which they are constantly exposed and new minerals are formed which are more stable under new conditions. The altered rock rapidly crumbles and disintegrates under the mechanical effects of erosion, and its constituents are transported by wind, water, or ice, and redeposited as sediments or remain in solution.

The main problem in geochemistry of sedimentation is the chemical break down of some minerals and the formation of others. The silicates are the most important, since they constitute more than 90% of the earth's crust (including quartz as a silicate). Recent investigations have shown that during the initial attack, a silicate mineral goes into ionic solution, and even the silica and the alumina are at least for a short time in true solution (Correns, 1949, p.208). At the surface of a crystal, unsatisfied valencies exist, which are the loci of reaction with water molecules. Hydration and hydrolysis follow, whereby strong bases, such as potassium, calcium, and magnesium, are removed and oxygen anions in the crystal lattice may be partly replaced by the hydroxyl ions. Aluminum and silicon attract OH ions strongly; aluminum probably groups six OH around it to assume its preferred six-fold coordination, whereas silicon remains in four-coordination. When first set free, these elements are in ionic solution, but the ions tend to aggregate and to form clusters of colloidal size. When first formed, these colloidal aggregates are probably amorphous, but later they arrange in definite crystal lattices, such as those of the clay minerals. Some silicate minerals may not undergo complete break down of the lattice during weathering; for example, biotite and muscovite may perhaps pass directly into clay minerals by ionic substitution, whereby fragments of the sheet structure may be directly incorporated in the new minerals. Therefore, relative stability of ions of different element in water is important. The most stable are the alkali metal ions, followed by those of the alkaline earths, and they are mostly carried away in solution. Silicon, aluminum and iron, on the other hand are generally soon redeposited as insoluble compounds and new minerals are formed from them at an early stage of weathering.

The Chemical Composition of Sedimentary Rocks

The chemical composition of sedimentary rocks is highly variable much more than that of igneous rocks, since sedimentation further causes diversification. However, it is considered that the average composition of sedimentary rocks should correspond fairly closely to that of igneous rocks, since all sedimentary rocks have been derived from igneous rocks by processes of weathering. The changes that occurs is due to loss of some elements such as sodium, which tends to accumulate in solution in the ocean, and the addition of components from the atmosphere and hydrosphere, such as oxygen, carbondioxide, and water.

The significant features of the chemical composition of sedimentary rocks are the dominance of potassium over sodium, alumina in excess of the 1:1 ratio to alkalies and calcium, high silica in sandy and cherty rocks, high lime and magnesia in the carbonates, and the presence of iron mainly in the ferric state.

The Mineralogical Composition of Sedimentary Rocks

The mineralogy of sedimentary rocks is characterized by two distinct types of material i.e. resistant minerals formed from the mechanical breakdown of the parent rocks and minerals newly formed from the products of chemical decomposition. The common and abundant minerals of sedimentary rocks are quartz, feldspar, calcite, dolomite and clay minerals. Of these, quartz and the feldspars are the abundant detrital minerals i.e.

those set free by the mechanical breakdown of the pre-existing rocks. Quartz is very resistant to chemical breakdown while feldspar is less resistant and they are chemically decomposed by prolonged weathering. Calcite is precipitated from solution either by physicochemical changes or by the vital processes of organisms. The geological evidence indicates that dolomite can form in sedimentary rocks from limestone as a product of diagenetic reaction from magnesium-rich waters like sea water.

Kaolinite, montmorillonite, illite, and chlorite, together with a number of less common species, constitute the clay mineral of sediments and sedimentary rocks. The clay minerals are stable secondary products formed by the decomposition of other alumino silicates. It is significant that all these new formed minerals have layer lattice structures, which seem to have greater stability than other types under surface conditions. They are all hydrous alumino silicates. In most sediments their average grain size is very small, generally less than 0.005 mm in diameter, and ranging down to colloidal dimensions, and as grain size diminishes, the perfection of crystallinity also decreases, and probably no sharp break exists between imperfectly crystallized clay minerals and amorphous material. Not only mechanical mixtures, but also "mixed" crystals in which molecular layers of more than one clay mineral species are interleaved in a single crystal, are often formed.

Physicochemical Factors in Sedimentation

The geochemistry of sedimentary processes is essentially the geochemisty of reactions taking place in the presence of water. Water is a unique solvent as it can dissolve number of substances and can hold them in solution.

Hydrogen-Ion-Concentration

The hydrogen-ion-concentration of natural waters is of great significance in chemical reactions accompanying sedimentary processes. In pure water at 20°C the hydrogen ion concentration is 10^{-7}moles/liter. If the concentration is greater than that of pure water at the same temperature, the solution is said to be acid and if it is less, then alkaline. The pH of the medium is particularly significant in controlling the precipitation of hydroxides from solution.

Oxidation-Reduction Potentials

Many elements are present in different oxidation states in the earth's crust. The commonest is iron, then manganese, sulphur, vanadium, copper, cobalt, nitrogen etc. The stability of an element in a particular oxidation state depends upon the energy change involved in adding or removing electrons. A quantitative measure for this energy change is provided by the factor known as oxidation-reduction or redox potential and can be shown by a reference standard $H_2 = 2H^+ + 2e$ (e=electron) (i.e., the removal of electron from hydrogen atoms or to hydrogen ions) (Mason, 1960).

Colloids and Colloidal Processes

The colloidal state is one of the fine state in which the size ranges from approximately 10^{-3} to 10^{-6}mm. The colloidal solution grade into true solutions on one hand and into suspension on the other without any distinct line of demarcation.

Colloidal particles are electronically charged and this charge may originate either by adsorption of ions from the liquid or by the direct ionization of material of the particle. Colloidal particles may be either crystalline or amorphous. Most inorganic colloids are crystalline. An important property of collidal particles is their ability to bind and concentrate certain substances through adsorption. Through these processes, many ions may be removed from natural waters. The clay minerals, especially montmorillonite, show a marked adsorption capacity and the chemical adsorption of potassium ions by montmorillonite may result in the formation of illite. The colloids adsorb heavy metal ions and water borne pollutants: hence the fates of reactive elements and of many pollutants in the environment depend to a large extent on the movement of colloids in aqueous systems. The colloids in natural waters are characterized by an extreme complexity and diversity: organism, biological debris, organic macro-molecules, various minerals, clays, an oxides, partially coated with organic matter. Many complex ions such as those containing arsenic and molybdenum and ions of the heavy metals are also adsorbed and thus removed from solution by natural colloids. Such elements are often enriched in sedimentary iron and manganese ores.

Products of Sedimentation

It is well established that weathering, erosion, and sedimentation leads to a marked separation of the major elements. Silica concentrates in the resistate sediments, alumina in the hydrolysates, iron and manganese in the oxidates, and calcium and magnesium in carbonates. Significant part of the sodium remains in solution and eventually accumulates in the ocean, while potassium is adsorbed by the clays and may form the minerals illite and glauconite. However, the fate of the minor elements during sedimentation has not been thoroughly investigated, and much less is known about their behaviour under these circumstances. Krauskopf (1955) evaluated the available data on the distribution of minor and trace metals in sedimentary rocks and concluded that nearly all the minor metals are somewhat more concentrated in shales than in limestones or sandstones, but the amount in ordinary shales is seldom more than twice the crustal abundance. The greatest enrichment of rare metals is in the black shales rich in organic metal, which may show marked enrichment in any or all of the elements Ag, Au, As, Mo, Ni and V. Sedimentary silicates, carbonates, and evaporates seldom show marked concentration of rare metals.

Lake Sediment Chemistry

Lakes are ideal place of deposition of weathered material derived from the surroundings and transported from distance sources through rivers, streams and nalas. It is a local site of sedimentation strongly influenced by its surroundings and bears records of events, which takes place in the surrounding drainage basin and are generally obliterated in the rocks. Lakes and their sediments change rapidly in response to rapid change in the drainage basin, even seasonal changes are recorded. The study of lake sediments particularly of core samples from depth helps in reconstructing depositional history (sedimentation), climatic changes, with time besides elemental pollution due to natural and anthropogenic sources. In fact, sedimentation in lakes is more affected by regional tectonic and climatic processes than the sedimentation in the oceans. However, many lake studies have shown that regional climatic oscillations can be related to global climatic processes. Investigation have revealed that climatic oscillations discernible from ice cores and high latitude sediment cores have their counterpart in the tropics (deGaridel-Thoron et al., 2001; Kundrass et al., 2001; Sirocko *et al.*, 1993, 1996). Early studies of lakes considered the sediments as only a repository having little or no additional reaction with the lake once deposited. In recent years role of sediments in the dynamics of lake system is realized. Recycling of mineralized organic matter, especially the nutrients, in sediments by organic decay and pore fluid transfer processes are now recognized as essential components of models that attempt to describe the nutrient dynamics of lake and reservoir systems (Allen and Kramer, 1972; Gahler, 1969; Lerman and Brunskill, 1971; Middlebrooks *et al.*, 1974). In addition, the active role lake sediments play in regulating cycles of trace metals, radionuclids, and synthetic organic chemicals is now gaining increasing attention. Similarly, the mineralogical composition of lake sediments and its importance in helping to understand lake processes is also gaining importance. It is realized that to properly elucidate the relationship between the constituent inorganic and organic phases of lake sediments and lake sedimentation processes, the mineralogical and chemical aspects of lake sediments must be elaborated. In fact, it is not always possible to separate mineralogical and chemical evidence in lake sediment studies.

Investigations of the mineralogical and chemical characteristics of sediment particles supplied to the lake by streams, shore erosion, atmospheric fall out, and cultural sources can reveal important information about the bedrock and surficial geology of the drainage basin, weathering, the relative importance of the various inputs, and quantitative estimates of cultural loadings on the lake systems. The latter is very important in the present times because, in about 100 to 150 years anthropogenic sources of inorganic and organic pollution in the lake system has increased considerably. This is due to increasing population, industrialization, deforestation, craze to produce more which entails reckless use of chemical fertilizers, pesticides, insecticides, manures etc. As the lakes are more or less confined water bodies, such activities are significantly recorded and needs to evolve proper planning to control hazards to human and aquatic life.

As the mineralogy of lake sediments is little altered by post depositional reactions i.e., diagenesis and that physical and biological mixing in sediments are small relative to the total amount of sediments deposited in the lake, the mineralogy and chemical variation of the sediments with depth can provide information on the physical and chemical changes that have occurred in the life of the lake. Such informations are found useful in evaluating long-term climatic changes of the region and in helping to assess the effects of cultural activities on the lake. In order to have a better understanding of sediment chemistry, and as the lake sediments play an active role in regulating the composition of lake systems, the factors involved in lake-sediment interaction need to be considered. Of particular importance are the effects of sediment grain size, pore fluid processes, the effects of biological mixing and sediment resuspension and variations in rates of sedimentation (Jones and Bowser, 1978).

Grain Size

One of the most important factors in determining mechanisms and rates of transfer of solutes between lake water and sediments is grain size. Investigation have shown that fine-grained materials have the most potential for interaction with lake water. Surface areas of typical clay-size sediment are measured in square meters per gram (Grim, 1968), whereas sandsize sediment areas are measured in tens of square centimeters per gram. It has been pointed out by Jenne (1976) that amorphous and crystalline ferric oxides commonly occur at surface coatings on grains they promote chemical activity even though their concentration is low. However, Gibbs (1977) suggested that thickest coatings of iron and manganese oxides on river-transported particles occur not on the smallest grains but on silt-size particles and that oxide coatings are inherited in the weathering zone where the rate of supply of metals to the grains is the controlling factor.

The detrital organic matter resulting from lake productivity is relatively fine-grained and low in density, thus preferentially accumulate within fine-grained silicate fraction. Because oxidation of organic matter is an essential factor in most sediment diagenesis (or authigenesis), such sediments are generally the most chemically reactive parts of the lake system.

Mineralogical differences in clays are also important in regulating solute transfer to lake waters because of the variations in surface area for different minerals of equivalent grain size and the significant difference as in ion-exchange and sorptive capacity of various clay mineral species encountered in lakes (Table 6.1).

Pore fluids

Chemical reaction between the organic and inorganic phases of lake sediments and fluids trapped within (or moving slowly through) the sediments is a significant process in most lake systems. The chemical gradients existing between the sediments and overlying lake water and physical processes of sediment compaction and bioturbation cause the transfer of water and solutes across the sediment-water interface. Thus, chemical interaction between lakes and sediments is largely regulated through solute transfer associated with pore fluid processes. The interrelationship between sediments and lakes is important especially with regard to the dynamics of nutrient elements (Jones and Bowser, 1978).

Table 6.1: Cation Exchange capacity and surface areas of Typical Clay Minerals

	Kaolinite	Illite	Chlorite	Smectite	Source/ Tech. vermiculite
Cation exchange capacity(meq/100g)	3-15	10-40	10-40	80-150	100-150
Surface area (m^2/g)	16	56	-	38	-
	12	52-82		203-250	-
	114		92	234	75
	57		97	221	87
	19		127	778	
	20		139	775	237
				760	548

(Taken from Jones and Bowser, 1978)

Biological Mixing (bioturbation) and Sediment Resuspension

Besides diffusional and advective transport of various solutes between lake water and sediments the role of sediment compaction, rapid sedimentation and groundwater has been emphasized (Berner, 1971; Lerman and Brunskile, 1971; Moore and Silver, 1975; Manheim, 1976). Also the important role of biological mixing or bioturbation, and/or sediments resuspension as a major solute transfer process in bottom sediments of organically productive waters such as lakes and coastal marine sediments has been realized (Schink and Guinasso, 1977; Goldhaber et al., 1977; Benninger *et al.*, 1977). Studies show that burrowing and ingesting organisms are present in a wide variety of lake sediment environments and can effectively mix sediments over dimensions of several centimeters over relatively short periods of time (Wetzel, 1975). The generally higher sedimentation rates and greater organic productivity of many lakes would indicate that mixing models should be particularly useful in lake sediment studies.

Sedimentation Rates

Systematic Studies of sedimentation rate variations are essential for interpretation of a variety of lake sediment interaction processes and also important in the study of the kinetics of transfer between the lake sediment reservoirs. Mass accumulation rates of minerals, but also organic debris, nutrient elements, and trace metals in sediments and their temporal variations require accurate sedimentation rate measurements. Studies of temporal variations in lacustrine sedimentation and the relation of such variations to climatic and/or anthropogenically induced changes in lakes, require knowledge of sedimentation rates and their temporal variations over time scale generally ranging from a few years to tens of thousands of years.

Various methods have been adopted for estimation of rate of sedimentation. In the latter part of the last century, the common technique used, in estimation of sedimentation rate is based on the study of rapid rise of ragweed pollen (Ambrosia) due to extensive forest cutting and soil cultivation by many (Bortleson and Lee, 1972; Craig, 1972; Kemp *et al.*, 1974; and Maher, 1977). However, the recent dating techniques involve natural radio-isotope lead-210 estimation (Krishnaswami et al., 1971; Koide et al., 1973) and the bomb fallout nuclide Cesium-137 (Pennington *et al.*, 1973). The relatively short lived isotopes (22.3 years for Pb^{210} and 30 years for Cs^{137}) are ideally suited for lake sediments whose rates of sedimentation are on the order of a few millimeters per year (for details see Krishnaswami and Lal, 1978).

Pb^{210} dating Method

The Pb^{210} dating technique is one of the most commonly used dating techniques of recent lacustrine sedimentary deposits. Many human impact and pollution studies rely on it to calculate recent increases of anthropogenic fluxes. 210 Pb is naturally occurring radioisotope in decay series of 238U and includes 222 Rn, which escapes into the atmosphere at a rate of about 42 atoms per minute per square centimeter of land surface. The radon atoms subsequently decay through a series of short-lived daughters to 210 Pb:

$$238U \rightarrow 226Ra \rightarrow 222Rn \rightarrow 210Pb \rightarrow 206Pb$$

Lead- 210 is rapidly removed from the atmosphere by meteoric precipitation and dry fall out, giving it an atmospheric residence half-life of only about 5-40 days (Appleby and Oldfield, 1992). After removal from the atmosphere, 210Pb is deposited in snow and ice of glaciers, in lakes and oceans, and in soil, from which it is absorbed by plants. In addition, 210Pb occurs in U-bearing minerals in secular equilibrium with 238U and is released into the environment as a result of weathering of such minerals.

The common assumptions of the Pb^{210} technique are: (i) Steady-state conditions exist in the sedimentary deposit with respect to the Pb^{210} flux and radioactive decay (half-life of 22.3 yrs); (ii) sediments can be considered as a closed system with respect to the sediment and Pb^{210} fluxes; (iii) the Pb^{210} has a short removal residence time in comparison with the residence time of the lake water, which means, that most of the Pb^{210} reaching the lake will be transferred efficiently to the sediments; (iv) no post depositional mobility of Pb^{210} occurs in the sediments; and (v) the excess Pb^{210} can be calculated from the total Pb^{210} and the Ra^{226} activity in the sediments, i.e. as the difference between the activities of the two nuclids, often it is assumed that the Ra^{226} activity is constant throughout the core, and therefore, that its activity can be determined from the constant Pb^{210} activity at depth, where Pb^{210} and Ra^{226} can be assumed to be in radioactive equilibrium. Pb^{210} enters a lake or reservoir either directly or through rain or indirectly in run-off from the catchment. A secondary source of Pb^{210} in the water column is from the decay of Rn^{222} in solution, which may or may not be maintain by Ra^{226}. Pb^{210} is deposited at the mud/water interface by sedimentation and the exchange processes, and becomes incorporated into the sediment column. Ra^{226} is also present in the sediment as part of the erosive input of particulate material from the catchment. Pb^{210} formed within the sediment by decay of Ra^{226} is referred to as supported Pb_{supp}^{210}, which is normally assumed to be in equilibrium with the Ra^{226}. Pb^{210} in excess of the supported activity is termed the unsupported Pb_{exe}^{210} and is considered to be derived predominantly from atmospheric input. This is measured as:

$$Pb_{exc}^{210} = Pb_{tot}^{210} - Pb_{supp}^{210}$$

Pb_{supp}^{210} is determined from Ra^{226}, which is in secular equilibrium with Pb^{210} (Krishnaswami and Lal, 1978). Since Ra^{226} is fairly constant throughout the length of the core, the activity of supported Pb^{210} is also assumed to be constant. The exponential decay of the atmospherically derived Pb_{exc}^{210} can be used to estimate the age of a sediment layer as long as it is significantly higher in activity than the Pb^{210} already present in the soil Pb_{supp}^{210}.

The vertical distribution of Pb_{exc}^{210} is studied from the equation-

$$Ax = Aoe^{-\lambda t} \qquad ......(1)$$

where Ax is excess Pb^{210} activity per unit weight of sample at depth x (cm) beneath the water-sediment interface,

t is age corresponding to depth 'X'

Ao is activity of excess Pb^{210} at sediment-water interface.

λ = decay constant (0.0311 yr^{-1}), therefore it can be shown as:

t = $1/\lambda \, In \, (A^0/A_x)$

Since annual rate of accumulation is constant such as b= x/t, where b is the annual rate of accumulation then:

$$In \, Ax = In \, A^o \, (\lambda - x/b)$$

A plot of In A as a function of depth (x) gives a curve, the slope of which is proportional to the sedimentation rate, i.e.

$$S = -\lambda/b$$

where, S = rate of sedimentation

and b = slope of the curve (Fig.6.1).

Cs^{137} dating Method

The use of radionuclids with relatively short half-life, such as Pb^{210} and Cs^{137}, to date recent sediment is well established (Krishnaswami and Lal, 1978). Cesium (Z=55) is an alkali metal in group IA of the periodic table. It has only one stable isotope (133Cs) and 35 unstable isotopes none of which occur naturally. Fourteen of the unstable isotopes are fission products (excluding isomers), and three of these have long half-lives: 134Cs (2.065 years), 135Cs (2.3 x 10^6 years), and 137Cs (30.17 years). Cesium-137 is an artificial radioisotope formed by nuclear fission and has a half-life of about 30.17 yrs (Lederer and others, 1967). These isotopes of Cs were released into the atmosphere by the testing of nuclear weapons in the atmosphere between 1945 and 1980. The short-lived fission-product isotopes of Cs, including 134Cs, decayed in less than about ten years after the last atmospheric test. Only 135Cs and 137Cs remain alive in soil, in recently deposited lacustrine and marine sediment, in continental ice sheets, and in the oceans. The rate of decay of 135 Cs is much lower than that of 137Cs because of its long half-life (2.3 x 10^6 years). Cesium-137 decays to stable 137Ba by emitting β-particles with a maximum energy of 0.514 MeV. In addition the product nucleus emits a γ -ray at 661.64 KeV, which permits the decay rate of 137Cs to be measured by γ -ray spectrometry.

The sources of 137Cs include not only fallout from the past testing of nuclear weapons but also the accidental explosion of the nuclear reactor at Chernobyl and the growing amount of spent nuclear fuel that continues to be held in temporary storage. This isotope has been introduced into the atmosphere in irregularly varying amounts since nuclear testing above ground began in 1945 (Durham and Joshi, 1980). Dating methods based on Cs-137 depends on the imprint of an irregular influx of wet and dry atmospheric deposition in sediment layers or its absence before 1945 (Krishnaswami and Lal, 1978). Cs-137 like Pb-210 is strongly bound by sediments and tends to remain trapped in a sediment layer, which is subsequently buried. Koide et al., (1973) have speculated that in normal coring devices, there is a possibility of losing the top portions of the deposit and have therefore suggested, that this loss may be checked by the presence of Cs-137.

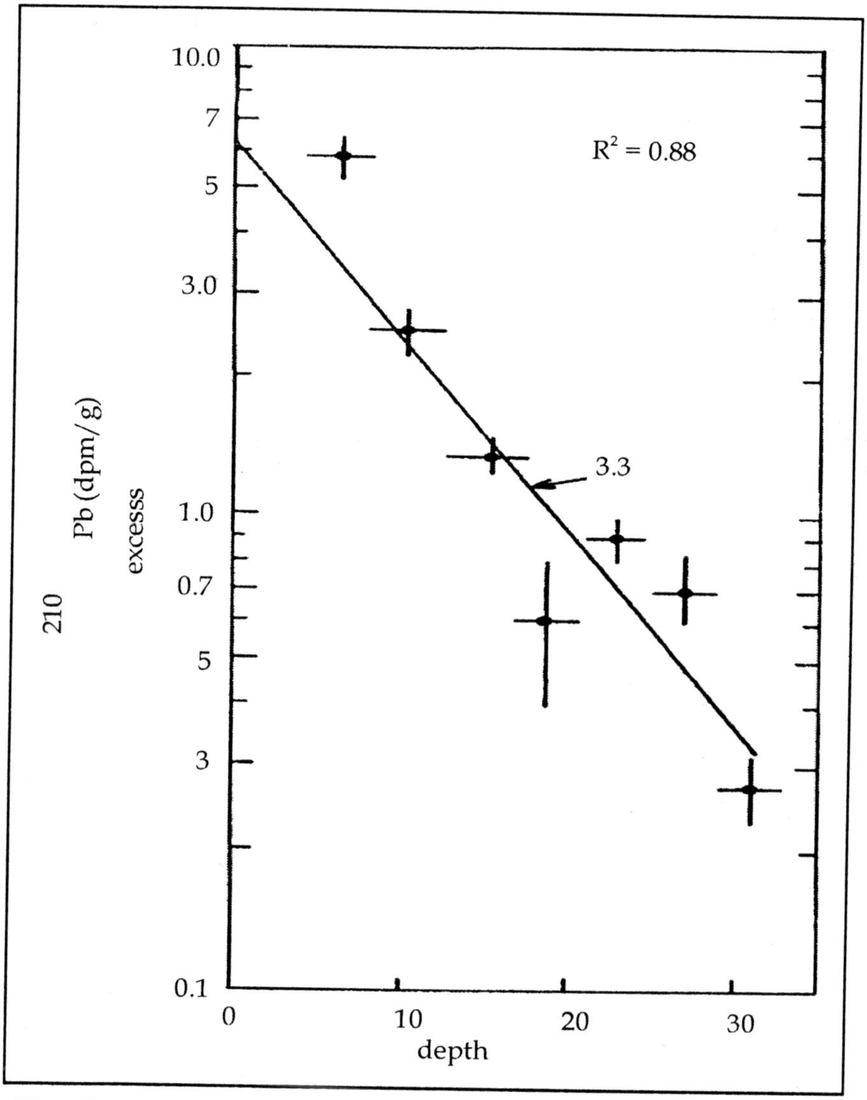

Fig. 6.1. Plot of excess Pb-210 activity (dpm/g) vs depth in sediment core from Renuka Lake. The measurement of the slope of the line gives rate of sedimentation.

The Cs-137 activity is determined by gamma counting 20g of the over-dried samples for up to 800 min using a 4in x 4in NaI detector-multichannel analyzer system. A Cs-137 standard, having essentially the same geometry and density is used. The limit of detection for Cs-137 by this method is 0.1 pci/g and the counting error is < ±10 percent in the upper layers of the core. Measurement is also made for this nuclide using a 30 ml lithium-drifted germanium detector coupled to a 4096 channel multichannel analyzer. This method permits a clear separation of the 661 Kev Cs-137 gamma rays from possible contributions from 208 Tl (583 Kev) and 214 Bi (609 Kev). The samples are counted for at least 24 hr to obtain good statistical accuracy.

Chemical Composition of Lake Sediments

The chemical composition of lake sediments may show considerable variation due to variation in rock types of the drainage basin/catchment area, weathering intensity and cultural activity. Therefore, it is difficult to draw any generalization about lake sediment compositions from the chemical data. Only a few general observations can be drawn from such data. The high calcium values would indicate a significant fraction of carbonate in sediment. The sediments analysis rich in calcite ($CaCO_3$) and low in Mg can reflect either allogenic or endogenic sources. The Ca/Mg ratios of sediments close to stoichiometric dolomite, presumably reflect actual detrital dolomite input. The silicon to aluminum ratios vary from low values of about 2 to close to 15 and in predominant clastic (allogenic) sediments such values would generally reflect variations in quartz to feldspar ratio. Feldspathic sediments would have Si/Al ratios of nearly 2 or 3, and more quartzose sediments will have much larger Si/Al ratios. The high correlation co-efficient = 0.95 of Si/Al ratios may suggest that both elements are from the allogenic (clastic) fraction of the sediment with little contribution from endogenic phases. In addition, considerable amorphous silica is contributed to lake sediments in productive lakes from diatoms. A more reliable picture would emerge, which dominantly control the lake sediment chemistry, when chemical data are interpreted in combination of the major constituents mineral phases.

The common major mineral phases of lacustrine sediments are quartz, feldspar, calcite, dolomite, ilmenite, smectite, apatite, and iron sulfides or iron oxides. The aluminosilicates are mostly allogenic minerals whereas carbonate (calcite) is endogenic, and authegenic minerals in lacustrine sediments are oxides, phosphates, and sulphides.

Among the trace elements, rare earth elements (REE) are the most useful as they have important applications in igneous, sedimentary and metamorphic petrology. The REE comprise the series of metals with atomic numbers 57 to 71 lanthanum (La) to lutetium (Lu). The low-atomic-number of the series are termed the light rare earths (LREE), those with higher atomic numbers the heavy rare-earths (HREE) and less commonly the middle members of the group, Sm to Ho, as middle REE (MREE).

Various studies have shown that the REE contents of most "Shales" are very similar, being enriched in the LREE relative to the HREE when normalized to chondrites. Rare earth element concentration in rocks are usually normalized to a common reference standard, which commonly comprises the values for chondritic meteorites. It has also been shown that REE contents of most sediments and sedimentary rocks are similar in the relative abundance of the individual elements although, they differ in absolute concentrations.

The recognition of the similarity in REE contents of sediments in general, and that of "shales" in particular, led to the idea that the average abundance of REE in sedimentary rocks could be taken to represent the average REE content of the crust. In other words sedimentary processes could be considered as causing homogenization of the REE fractionation, which occurs during the formation of igneous rocks.

During weathering and erosion only very minor amounts of REE goes into solution, for only a few percent of the REE entering to oceans are dissolved (Martin et al., 1976). The bulk of the REE in eroded material is contained in clays. The low concentration of REE associated with quartz, the other major silicates present in eroded material, are reflected by the fact that sandstones have lower REE contents than "shales". The REE in sandstones are probably largely contained in the clays which are present in them. However, if detrital grains of minerals such as apatite and zircon which concentrate REE in them by igneous process, are present in abundance in a sandstone, they make the sandstone rich in REE. Similarly, high concentration of the REE in the detrital fraction of transported material will indicate that the REE pass from weathering to deposition almost exclusively without taking part in any significant chemical processes. This helps in delineating source rock.

The REE contents of sediments and sedimentary rocks reflects the mineral contents of these deposits, and hence the processes by which the minerals formed and were incorporated into the deposits. Shales generally contain similar absolute REE abundances and have their REE fractionated relative to another. This characteristic of shales generally indicate that fine grained sedimentary material is efficiently mixed during weathering, erosion and transportation. The REE abundances of shales are, therefore, considered to represent the REE contents of the upper continental crust. The individual REE contents in sediments and sedimentary rocks, other than those of alumino-silicate mud rocks, are generally fractionated relative to one another as those of shales. The absolute REE abundances in these sediments are not only less than those in shales but decreases in the order of grey wackes, sandstones and limestones (Haskin *et al.*, 1966). This decrease in REE concentration presumably reflects the decreasing clay mineral and rock fragment contents and the increasing quartz and biogenic carbonate contents of these rocks, since the former constituents will contain higher REE contents than the latter ones. The available data, suggest that the REE are unaffected by diagenesis (Chaudhuri and Cullers, 1979). Several studies have also shown that metamorphism has not altered REE content in Precambrian sediments which are essentially the same as younger and present-day sediments (Wilderman and Condie, 1973; Wilderman and Haskin, 1973). These generalization can be understood better by discussing specific cases of different lake sediments.

□□□

Chapter 7

SEDIMENT CHEMISTRY EXAMPLES

As mentioned above, geochemical data of sediments and sedimentary rocks are very few. It has been found that the chemistry of lake sediments reflect sediment sources, provenance, weathering in source area, recycling, fractionation of sediments and climatic changes. Because of these, lacustrine sediment studies are gaining importance in recent years. In order to understand these, and to assess a reliability of the geochemical signatures preserved in the sediments, a few specific cases of lakes from, Himalaya are discussed.

Rewalsar Lake Sediments

The lake is situated in middle Siwalik Formation of lesser Himalaya and comprises of sandstone with silt-shale interlayer. The Siwalik Group (Middle Miocene to Lower Pleistocene) have been deposited in the foreland basin along the margin of the Himalayan collision mountain belt, south of Main Central Thrust (MCT) (Fig. 7.1) The sedimentary pile of these basins consists primarily of detritus eroded from orogenic belt and records the unroofing history of the orogen, the patterns of subsidence and nature of sediment dispersal across the basin. Thus, these deposits bear an important record of orogenic history and processes (Covey, 1986).

The major elements content in sediment samples show variation in SiO_2 and Al_2O_3 and corresponding increase and decrease in other oxides (Table 7.1), (Das and Haake, 2003). The high SiO_2 and low Al_2O_3 in some samples indicate high quartz content and coarser grain size of the detritus, whereas low silica and high Al_2O_3 represent enrichment of clay and finer grain sediments. It is often debated that weathered material (detritus) may not truly represent parent rock composition. The plot of log values (SiO2/ Al_2O_3) vs log (Fe_2O_3/K_2O) which are almost immobile elements, demarcate the catchment lithology of Rewalsar basin, remarkably similar to that outcropping in the lake catchment area (Fig.7.2) (Pettijohn *et al.*, 1972; Herron, 1988). This may strengthen the idea to review the chemistry of Rewalsar detritus and to evaluate features mentioned above which are presumably preserved faithfully as geochemical signatures in them. A negative correlation of major elements with SiO_2 indicates different rock sources. A striking depletion of CaO and Na_2O compared to Post Archaean Australian Shale (PAAS) indicate moderate to strong weathering due to their high mobility (Nesbitt et al., 1980; Wronkiewicz and Condie, 1987; Condie, 1993). The very low Na_2O/K_2O ratio indicates immaturity of sediments. This corresponds to petrographic study of middle Siwalik Group of sandstones (Tandon, 1972).

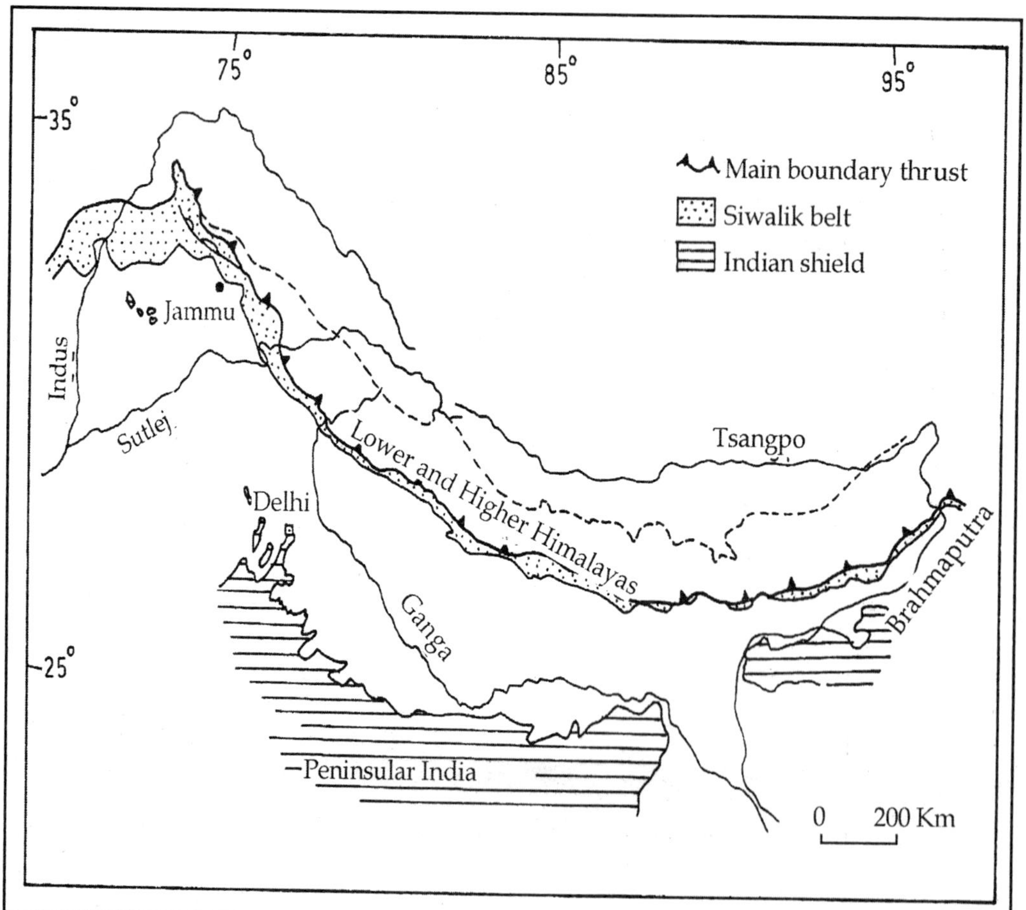

Fig. 7.1. Map of Himalaya showing foreland basin where Siwalik sediments were deposited.

Table 7.1: Major and Trace Element Data for the Sediment Samples of the Rewalsar Lake

Sample No.	Rew- 1	Rew-2	Rew-3	Rew-4	Rew-5	Rew-6	Rew-7	Rew-8
Major Oxides (Wt %)								
SiO_2	73.11	55.03	61.14	63.34	64.72	72.16	58.12	68.04
TiO_2	0.38	0.48	0.56	0.53	0.65	0.45	0.70	0.47
Al_2O_3	9.93	13.05	15.04	13.40	15.21	9.44	16.11	11.43
Fe_2O_3	2.77	3.85	4.55	4.11	4.20	2.58	6.07	2.77
MnO	0.05	0.09	0.05	0.06	0.05	0.04	0.08	0.06
MgO	1.22	1.47	1.57	1.51	1.79	1.29	1.87	1.43
CaO	0.33	1.44	0.55	0.52	0.42	0.26	0.55	0.39
Na_2O	0.53	0.47	0.53	0.55	0.62	0.54	0.50	0.45
K_2O	1.73	2.12	2.65	2.42	2.84	1.78	3.35	2.01
P_2O_5	0.10	0.19	0.12	0.11	0.09	0.10	0.20	0.11
LOI	9.39	20.13	11.33	12.82	8.24	10.97	11.86	12.72
Total	99.53	98.30	98.09	99.37	98.83	99.62	99.40	99.86
CIA	75.19	78.05	76.96	75.29	76.02	74.19	75.60	76.19
CIW	87.39	79.50	89.09	87.92	89.76	87.62	89.77	88.89
Trace Elements (ppm)								
Sr	25.8	50.30	40.9	40.3	40.1	25.6	51.7	25.0
Y	17.4	22.8	24.5	24.8	25.9	17.7	30.5	16.30
Zr	213.00	211.00	200.00	209.00	158.00	247.00	183.00	184.00
V	47.00	65.00	75.00	65.00	82.00	49.00	103.00	55.00
Cr	25.00	166.00	68.00	52.00	68.00	31.00	75.00	42.00
Ni	26.00	55.00	43.00	38.00	38.00	25.00	44.00	28.00
Cu	24.00	34.00	25.00	27.00	27.00	19.00	42.00	22.00
Zn	70.00	112.00	109.00	94.00	92.00	61.00	353.00	66.00
Rb	62.00	83.00	101.00	93.00	101.00	62.00	145.00	63.00
Nb	7.53	9.22	10.9	10.6	7.3	7.9	13.3	7.58
Pb	19.20	28.8	24.3	27.3	26.5	19.2	39.4	21.2
Th	9.81	13.00	13.50	13.70	14.10	9.82	17.8	9.52
U	2.09	3.00	2.89	2.82	2.92	2.16	3.34	2.08
Ba	232.00	369.00	437.00	402.00	457.00	245.00	564.00	255.00
Hf	6.10	6.28	6.19	6.41	4.99	7.32	5.61	5.55
Ta	0.735	0.941	1.048	1.039	0.926	0.768	1.204	7.36
Th/U	4.69	4.33	4.67	4.86	4.83	4.54	5.33	4.58
Zr/Y	12.24	9.25	8.16	8.43	6.10	13.95	6.00	11.29
Zr/Nb	28.29	22.88	18.35	19.72	21.64	31.26	13.76	24.27
Zr/Hf	34.92	33.59	32.31	32.60	31.66	33.74	32.62	33.15
Th/Zr	0.05	0.06	0.07	0.06	0.09	0.04	0.09	0.05

Contd...

Table 7.1. Contd...

K/Rb	231.54	210.95	217.71	215.92	233.32	238.23	191.71	264.75
Cr/Ni	0.96	3.02	1.58	1.37	1.79	1.24	1.70	1.50
Cr/V	0.53	2.55	0.91	0.80	0.83	0.63	0.72	0.76
Cr/Th	2.55	12.77	5.03	3.79	4.82	3.15	4.21	4.41
La/Th	2.507	2.392	2.377	2.313	2.461	2.485	2.337	2.563
Ce/Y	2.931	2.838	2.779	2.689	2.784	2.842	2.862	3.098
Zr/Yb	109.79	86.12	74.07	80.07	57.45	122.88	55.79	99.46
Zr/Th	21.71	16.23	14.81	15.25	11.20	25.15	10.28	19.32
K_2O/Na_2O	3.26	4.49	5.00	4.40	4.58	3.29	6.70	4.46
SiO_2/Al_2O_3	7.36	4.21	4.06	4.72	4.25	7.64	3.60	5.95
Al_2O_3/SiO_2	0.13	0.23	0.24	0.21	0.23	0.13	0.27	0.16
Al_2O_3/Na_2O+CaO	11.55	6.83	13.92	12.52	14.62	11.83	15.34	13.61

The quantitative estimation of weathering is commonly done by using formula:

$$Al_2O_3/(Al_2O_3 + CaO^* + Na_2O + K_2O) \times 100$$

where CaO* is the amount of CaO incorporated in the silicate fractions of the rock as proposed by Nesbitt and Young (1982), and known as Chemical Index of Alteration (CIA). This value ranges between 75.2 to 78.1 for Rewalsar Lake sediments and that for average shale from 70 to 75, which is close to the values for muscovite, illite and smectite, thus indicating large proportion of clay minerals. Unweathered igneous rocks show a value of about 50 whereas intensely weathered residual rocks forming kaolinite and gibbsite have values approaching 100. According to Nesbitt and Young (1982), the CIA values are not significantly changed due to metamorphism, therefore, even though source rock may be metamorphic or sedimentary, CIA value can be compared to shale standard and hence may reflect magnitude of weathering in the source area.

This value for PAAS is 69, indicating relatively higher degree of chemical weathering at Rewalsar source area.

The high field strength elements Zr, Hf, Y, Nb, Th, and U are considered to indicate provenance composition as they are immobile in their behaviour

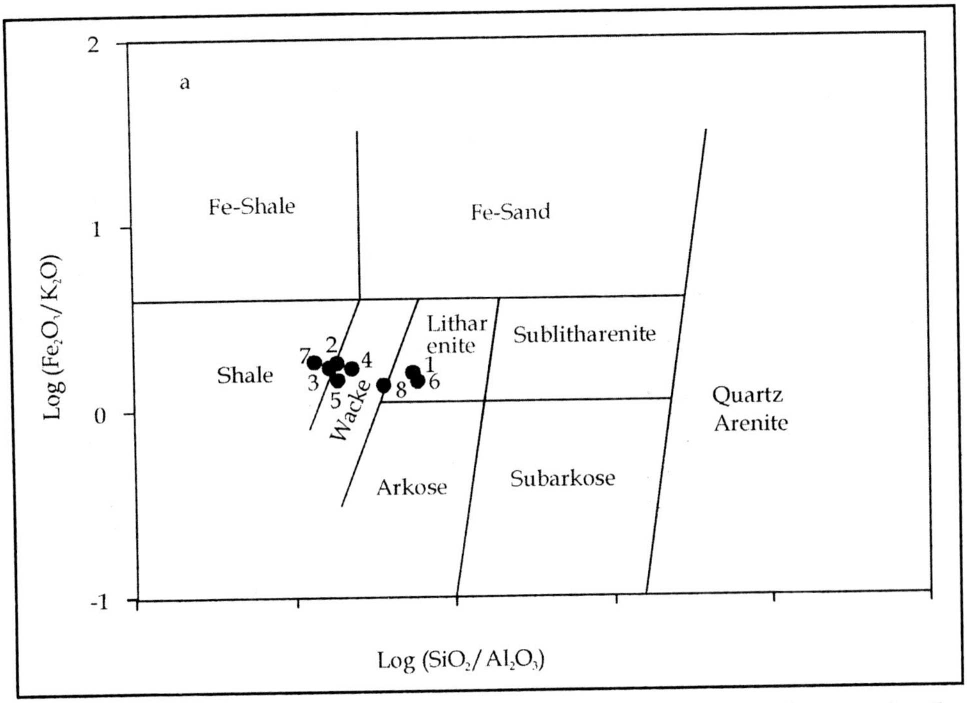

Fig. 7.2 (a). Major oxides log values plot on Herron (1988) diagram demarcating the rock types outcropping in the catchment of Rewalsar Lake.

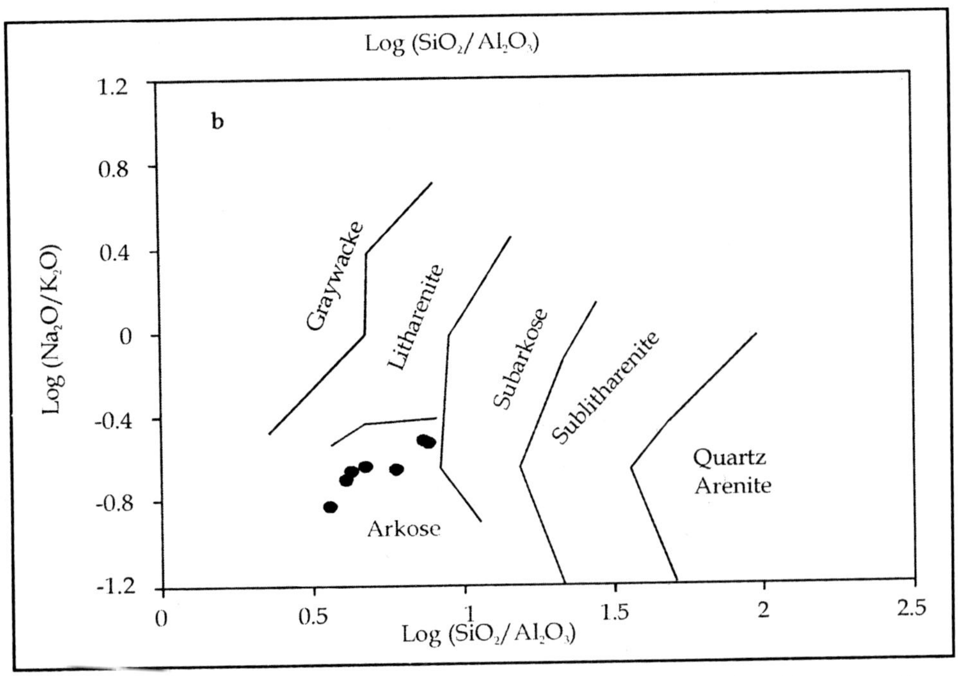

Fig. 7.2 (b). Major oxides log value plots on Pettijohn *et al.* (1972) diagram falls in Arkose field.

(Taylor and McLennan, 1985). Zr is enriched in silica rich sediments compared to the associated shales, as it has a tendency to preferentially concentrate in coarser sediments during weathering as detrital mineral. The high Zr/Th and Zr/Yb ratios in quartz rich samples of Rewalsar sediments support the above observation. The Zr and Hf show high positive correlation (r=098) and the Zr/Hf values range between 31 to 40 suggesting that these elements are controlled by zircons as these are nearly identical to those reported by Murali *et al.*, (1983) from the analysis of zircon crystals. However, its negative correlation with Y (r = -0.47) and HREEs suggests that the latter is not entirely controlled by Zr abundance. A negative correlation of Y with SiO_2 also indicates lack of zircon control in the distribution of elements.

The large ion lithophile elements (LILE) Rb, Ba, Sr and U positively correlate with Al (r=0.90,0.90,0.83 and 0.91, respectively), but Th shows a moderate value (r=0.58) indicating their distribution controlled mainly by phyllosilicates. The high linear correlation of Rb with K (r=0.97) indicate its association with illitic phases as Rb with respect to K is preferentially retained in weathered illite (Garrels and Christ, 1965). The small cations such as Na, Ca and Sr, being smaller in size, are selectively leached and removed from weathering profiles, hence depleted in sediments, whereas cations with relatively large ionic radii, such as K, Cs, Rb, and Ba may be fixed by preferential exchange and adsorption on clays and are represented in Rewalsar Lake sediments. This has also been suggested by Nesbitt et al., (1980); Wronkiewicz and Condie (1987) and Condie (1993). The low abundance of Sr is due to low Ca in the detritus because bulk of Sr reside in calcite due to its chemical affinity with the Ca. In magmatic rocks, it occurs commonly in plagioclase feldspar.

The REEs of sediment samples are normalized to PAAS (rock standard) and plotted graphically to understand source area composition (Fig. 7.3). The pattern shows a pronounced negative Eu anomaly typical of shales derived from cratonic source (Taylor and McLennan, 1981). Although the pattern of individual samples of Rewalsar are the same, but the concentration distinctly varies, separating into three distinct sets, indicating grain size control corresponding to the rock types, shale, wacke, and litharenite, outcropping in the catchment area and are thus consistent with the variation in major, trace and REE chemistry of detritus. This may suggest poor mixing of sediments on one hand, and REE concentration mineralogically and grain size controlled on the other.

The chondrite-normalized samples show LREE enrichment (La_N/Sm_N >3.07, and La_N/Yb_N > 8.45 ratio) and HREE (av. Gd_N/Yb_N 1.52) depletion pattern with steep negative Eu anomaly similar to UCC, PAAS and NASC (Nance and Taylor, 1976; Fleet, 1984) (Table 7.2; Fig. 7.3,b). The pattern resembles felsic igneous source or reworked sedimentary source. The slightly depleted REE contents in lake sediments compared to PAAS indicate weaker recycling process at source than in the latter, and that the heavy mineral content is deficient than in the felsic source rock. A high positive correlation between Al_2O_3 $(Gd/Yb)_C$ and negative correlation with SiO_2 indicate its association with clay rich fractions. In addition, the moderate negative correlation between Zr-$(Gd/Yb)_C$ indicates that the HREE fractionation is not solely controlled by the occurrence of zircon, which is also evident from the low negative correlation between Zr-HREEs.

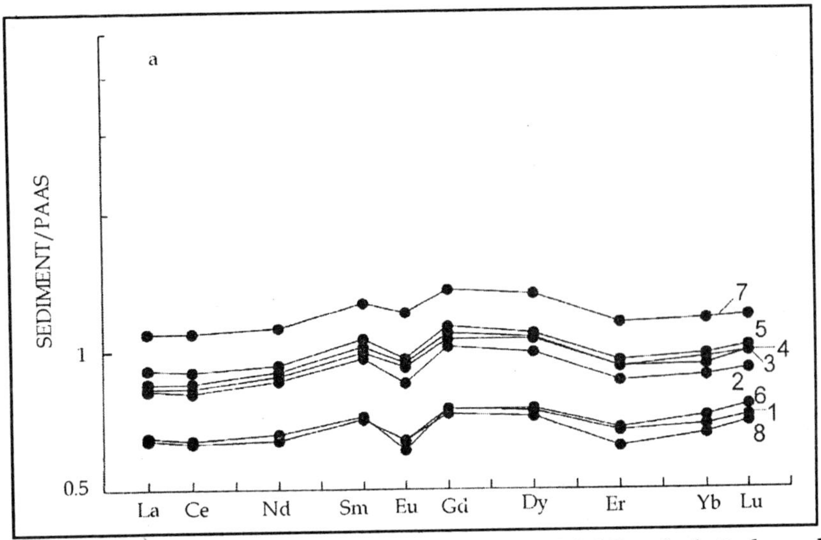

Fig. 7.3 (a). The REE of sediment samples normalized to PAAS and plotted graphically to understand source area composition. The pattern is the same but the concentration varies according to grain size.

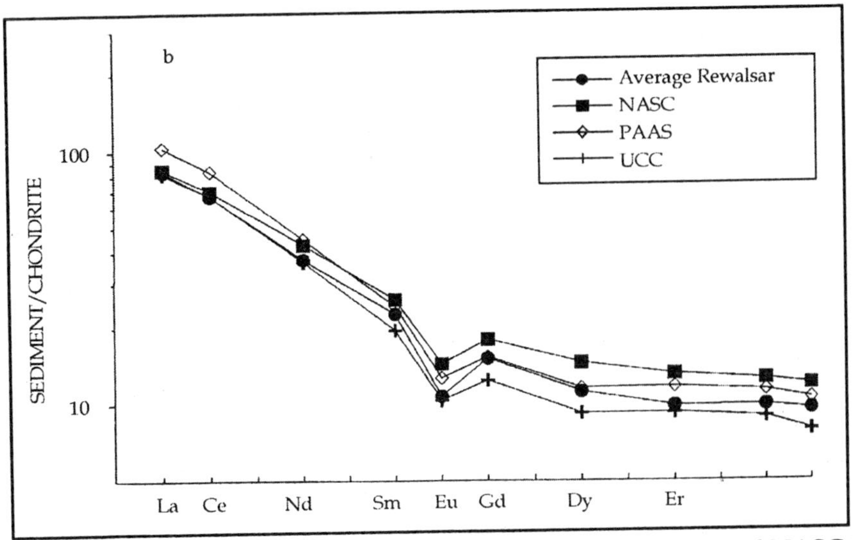

Fig. 7.3 (b). The chondrite normalized Rewalsar Sediment samples and NASC, PAAS, and UCC are plotted graphically, they show similar pattern with LREE enrichment, HREE depletion and steep Eu anomaly.

The REE patterns have also been used to infer source of sedimentary rocks, as the basic rocks contain low LREE/HREE ratio and no Eu anomalies, whereas more silicic rocks usually contain higher LREE/HREE ratios and negative Eu anomalies (Cullers and Graf, 1983). This suggests that the REE pattern of the source rocks is preserved in sedimentary rocks (Taylor and McLennan, 1985; Wronkiewicz and Condie, 1987, 1989). In the lake sediment LREEs are enriched and the HREE patterns are almost flat, with $(Gd/Yb)_N$ ratio varying from 1.41 to 1.58 with a negative Eu anomaly. Such patterns may indicate felsic source and a negative Eu anomaly may be regarded as an evidence of differentiated silicic source, similar to granite (McLennan, 1984; McLennan *et al.*, 1993; Taylor and McLennon, 1985, 1995).

Table7.2: Rare Earth Element (REE) data (ppm) for the sediment Samples of Rewalsar Lake.

Sample No.	Rew-1	Rew-2	Rew-3	Rew-4	Rew-5	Rew-6	Rew-7	Rew-8
La	24.60	31.1	32.10	31.7	34.7	24.4	41.6	24.4
Ce	51.00	64.7	68.1	66.7	72.1	50.3	87.3	50.5
Pr	5.73	7.29	7.66	7.52	8.04	5.61	9.67	5.57
Nd	21.10	27.4	28.70	28.0	29.90	20.5	35.90	20.60
Sm	4.03	5.42	5.71	5.58	5.94	4.00	7.13	3.99
Eu	0.676	0.939	1.03	1.01	1.06	0.706	1.33	0.706
Gd	3.54	4.80	5.10	5.00	5.29	3.50	6.38	3.45
Tb	0.554	0.674	0.763	0.740	0.811	0.553	1.023	0.546
Dy	3.25	4.34	4.69	4.66	4.82	3.30	5.87	3.17
Ho	0.666	0.863	0.973	0.952	1.02	0.68	1.17	0.632
Er	1.94	2.48	2.67	2.66	2.76	1.97	3.32	1.81
Tm	0.280	0.356	0.390	0.384	0.399	0.293	0.475	0.265
Yb	1.94	2.45	2.70	2.61	2.75	2.01	3.28	1.85
Lu	0.288	0.365	0.399	0.394	0.408	0.305	0.477	0.281
ΣREE	119.594	155.177	160.985	157.91	169.998	118.121	204.925	117.769
$(La/Yb)_N$	8.569	8.578	8.034	8.207	8.527	8.203	8.570	8.913
$(La/Sm)_N$	3.842	3.612	3.538	3.575	3.677	3.839	3.672	3.849
Ce/Ce*	1.043	1.037	1.053	1.003	1.044	1.044	1.054	1.047
Eu/Eu*	0.547	0.562	0.583	0.584	0.578	0.577	0.603	0.581
$(Ce/Yb)_N$	6.813	6.843	6.536	6.337	6.795	6.484	6.897	7.074
$(Ce/Sm)_N$	3.055	2.881	2.879	2.761	2.930	3.035	2.955	3.055
Gd_N/Yb_N	1.479	1.588	1.531	1.552	1.559	1.411	1.576	1.511
ΣLREE	106.46	135.91	142.27	139.50	150.68	104.81	181.60	105.06
ΣHREE	12.458	16.326	17.685	17.400	18.258	12.611	21.995	12.003
ΣLREE/ ΣHREE	8.545	8.324	8.045	8.017	8.253	8.311	8.256	8.753

However, Condie (1991) and McLennan and Taylor (1991) have emphasized that shales derived from cratonic settings have Eu anomaly ranging from 0.60 to 0.70, and the values of Rewalsar Lake sediments are close to this range. A high positive correlation between Th and LREE exist. The finer sediments show a little higher value than sands. As in sands La and Th increase, as grain size decrease, a grain size control is well reflected in a linear plot of La and Th (Fig 7.4a).

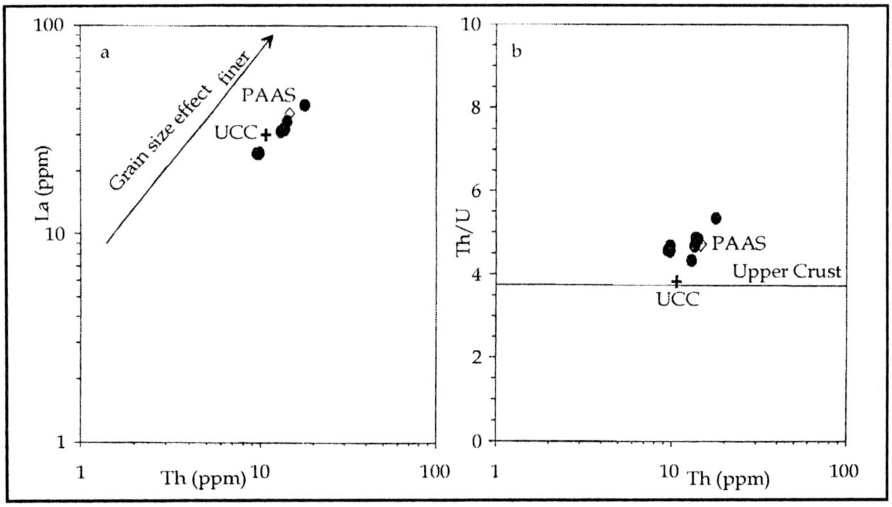

Fig. 7.4 (a). The La/Th plot of Rewalsar sediments lie between UCC and PAAS showing a common source and a linear trend indicating grain size sorting. **(b)** The Th/U vs Th plot of lake sediments indicate chemical weathering during which there is a tendency for an elevation of Th/U ratio above UCC.

Weathering trend is assessed from a Th/U vs Th plot. During weathering, there is a tendency for an elevation of Th/U above upper crustal igneous values of 3.5 to 4.0 (McLennan et al., 1993) (Fig. 7.4b). Rewalsar detritus show Th/U ratios ranging from 4.33 to 5.33, which are higher than upper crustal values (UCC) and close to PAAS (4.7), indicating moderate to strong chemical weathering at source area.

Positive linear correlation between Al_2O_3 and light and heavy REEs (r=0.932 and 0.877, respectively), suggest that REEs are dominantly contained in clay minerals (Wronkiewicz and Condie, 1987; Culler *et al.*, 1987). A high positive correlation (> 0.9) between REEs and major oxides suggest that the former is having a common source, other than the heavy minerals (Gromet and Silver, 1983; Bea 1996). Mica, biotite, chlorite, clay minerals constituting the sand-silt-shale fractions of sediments are thus the promising source and host REEs (Bauluz *et al.*, 2000; Taylor and McLennan, 1985). It is therefore, envisaged that the middle Siwalik rocks predominantly consists of lithic metamorphic rock fragments which are derived from lesser Himalayan domain (Tandon, 1972).

In a plot of log K_2O/Na_2O vs SiO_2 (%), Rewalsar sediments fall in a passive margin (PM) domain indicating that Rewalsar detritus (derived from middle Siwalik Group) have been deposited in a passive margin setting (Fig. 7.5 a, after Roser and Korsch, 1986). Discriminant function analysis has been used extensively in the investigation of relationships in the chemistry of the major and trace elements and also to discriminate between various tectonic environments and sedimentary provenance (Roser and Korsch, 1988). However, recycled sediment and those from a mixed source are much more difficult to interpret. The analysed samples plot in P 4 field of this diagram (Fig. 7.5b) indicating recycled orogenic terrain (quartzose sedimentary provenance source) and deposited in a passive margin regime. This is consistent with the immobile trace element chemistry mentioned above.

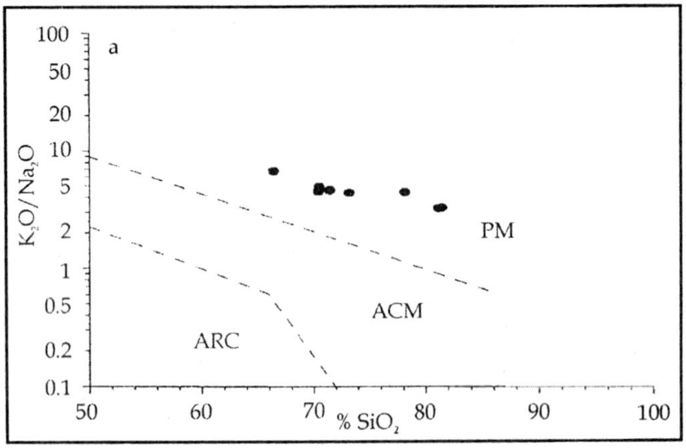

Fig. 7.5 (a). In a log K_2O/Na_2O vs SiO_2 plot Rewalsar sediments fall in passive margin (PM) setting.

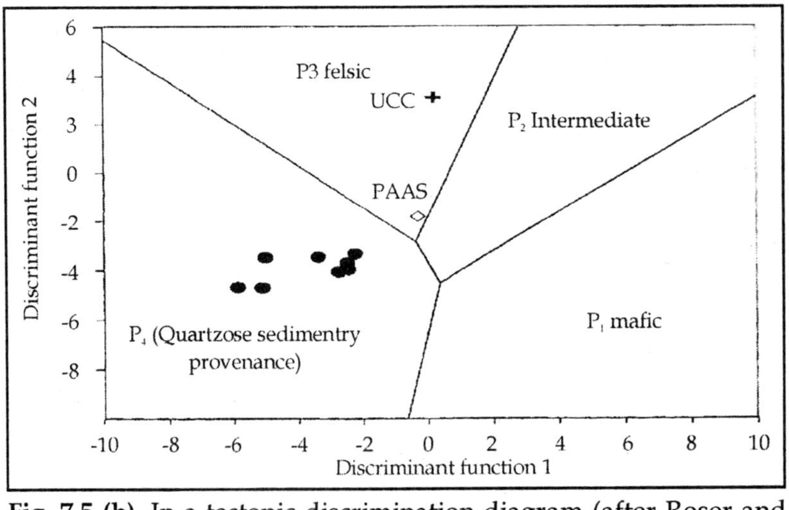

Fig. 7.5 (b). In a tectonic discrimination diagram (after Roser and Korsch, 1988), the Rewalsar sediments plot in P4 domain demonstrating that they are derived from cartonic interior or quartzose sedimentary provenance source.

Renuka Lake Sediments

The major element distribution reflects the mineralogy of rocks. For instant, sandstone have higher SiO_2 and correspondingly lower Al_2O_3 than shales due to their high quartz content. Similarly, most of the shales are enriched in K_2O, Fe_2O_3 and TiO_2 in comparison to sandstones, reflecting their higher content in clay phases and other clay-size grains. Some wackes are enriched in CaO and MgO due to occurrence of diagenetic calcite and dolomite minerals. The limestone will be enriched in CaO and will be very poor in SiO_2, Al_2O_3 and other major oxides.

The analytical data of Renuka sediments were recalculated on an anhydrous basis to 100% to evaluate the behaviour of major elements as the sediments are enriched in carbonate content (Table 7.3). The concentration of major elements thus derived are plotted against Al_2O_3 (Fig.7.6). Fe_2O_3, K_2O, TiO_2 and Na_2O increases from lake to wetland while P_2O_5 decreases indicating organic source of C_{ORG} (7.76-12.95%) in lake compared to wetland (1.64-3.49%). This indicates higher clay fraction in wetland sediments relative to lake sediments. The correlation values of Fe_2O_3, K_2O and TiO_2 with Al_2O_3 are positive (r=0.735,0.863,0.632 and r=0.839,0.908, 0.364 for the lake and wetland sediments respectively), suggesting common source. However, increasing trend of these oxides, indicate enrichment of phyllosilicate in wetland due to retention of finer particles in soil. The CaO and MgO, do not show any trend but enrichment in lake sediments, indicate their source from the limestone and dolomite of the lake catchment area. The depletion of major oxides except Ca and Mg, in lake sediments, in comparison to wetland having similar source rocks in the catchment and same weathering intensity, indicates SiO_2 dilution by CaO, and other oxides redistributed accordingly.

The major oxides ratio have been used by Pettijohn *et al.*, (1972) and Herron (1988) to distinguish rock types of an area/basin. The SiO_2/Al_2O_3 ratio reflect the abundance of quartz as well as clay and feldspar content (Potter, 1978), the Na_2O/K_2O ratio is an index of chemical maturity (Pettijohn et al.,1972), and Fe_2O_3/K_2O ratio is a measure of mineral stability as the ferromagnesian minerals tend to be amongst the least stable minerals during weathering. In a Pettijohn et al., (1972) diagram, wetland and some lake samples plot in arkose and some in litharenite fields (Fig.7.7a). The shift of the samples towards the arkosic field is due to high K_2O content, relative to low Na_2O in wetland samples, which decrease in Na_2O/K_2O ratio bringing the samples down to lie in arkosic field. In Herron's diagram (Fig. 7.7b), on the other hand, samples plot in the shale field and some wetland samples lie on wacke-shale junction due to high Fe_2O_3/K_2O ratio together with low SiO_2/Al_2O_3 ratio in them. This suggests shale dominated lithology surrounding the

Table 7.3: Major and trace element data of the Renuka Lake and Wetland Sediment Samples (LoI Free)

Sample No.	Lake							Wetland						
	GRI	GRII	GRII	GR IV	GR V	GR VI	GRVII	GRVIII	GR IX	GR X	NGRI	NGRIV	NGRV	NGRVI
Major Oxdise (wt%)														
SiO_2	56.04	49.17	48.72	51.31	47.98	50.71	51.54	63.87	65.49	59.08	64.98	62.50	65.16	65.99
TiO_2	0.37	0.46	0.41	0.42	0.38	0.38	0.52	0.75	0.78	0.75	0.87	0.83	0.81	0.84
Al_2O_3	21.13	14.32	19.87	13.42	13.44	13.05	15.10	16.53	16.39	19.64	17.12	18.92	16.97	16.75
Fe_2O_3	7.01	4.37	4.32	4.69	3.81	3.65	5.11	6.44	6.95	7.34	6.93	6.93	6.59	6.58
MnO	0.05	0.12	0.10	0.10	0.10	0.14	0.16	0.07	0.12	0.14	0.15	0.16	0.06	0.07
MgO	3.17	8.37	2.21	7.34	9.96	7.64	8.54	3.86	3.54	4.46	4.35	3.09	4.16	3.70
CaO	7.48	20.20	21.64	19.70	21.82	21.88	20.27	4.03	2.11	3.74	0.76	2.54	1.26	1.31
Na_2O	0.60	0.71	0.42	0.61	0.61	0.59	0.74	0.95	0.92	0.61	0.95	0.54	0.80	1.00
K_2O	3.62	0.82	2.12	1.86	1.39	1.49	1.89	3.17	3.30	4.01	3.81	4.31	3.88	3.47
P_2O_5	0.21	0.47	0.19	0.55	0.48	0.48	0.61	0.31	0.36	0.22	0.20	0.18	0.36	0.27
LOI	25.31	29.77	27.04	27.31	28.13	27.39	34.79	16.07	16.05	12.13	7.68	10.61	11.42	12.70
Total	99.68	99.01	100.00	100.00	99.97	100.01	104.48	99.98	99.96	99.99	100.12	100.00	99.99	99.98
CIA (%)	78.17	76.95	84.35	76.95	79.19	78.55	77.03	71.57	71.21	75.52	70.77	74.57	71.17	77.32
Trace element														
Ba	149	<5	51	<5	<5	<5	<5	113	130	168	297	361	251	241
Cr	38	12	5	11	5	4	13	49	55	43	81	57	67	72
V	87	52	64	56	54	48	51	84	89	102	115	115	100	103
Ni	36	37	32	34	32	32	32	41	42	49	50	44	46	45
Cu	33	43	32	51	39	44	30	32	32	40	37	38	39	33

Table 7.3. Contd...

Zn	198	91	101	128	93	102	97	268	234	164	224	141	187	275
Pb	39.2	30.9	36.3	33.3	30.7	28.2	22.5	35.9	42.8	58.9	44.2	52.2	39.4	40.7
Th	14.7	9.76	12.2	n.d	n.d	n.d	9.36	n.d	n.d	n.d	17	19.4	16	16.60
Rb	151	80	128	81	77	79	85	134	139	183	150	190	154	145
Sr	260	243	916	250	240	306	346	134	86	93	82	75	61	83
Y	23.9	16.9	<5	17.3	16.2	15.1	15.4	29.5	33.3	37.9	35.9	39.0	35.2	33.9
Zr	126	90	137	69	59	76	91	205	216	195	218	244	200	231
Nb	10.1	8.5	8.9	7.7	7.7	8.3	8.8	13.4	13.8	13.4	14.5	14.8	13.8	14.3
Hf	4.51	3.69	3.37	n.d	n.d	n.d	3.58	n.d	n.d	n.d	6.01	7.06	6.56	7.24
Ta	0.96	0.72	0.69	n.d	n.d	n.d	0.75	n.d	n.d	n.d	1.39	1.44	1.38	1.40
U	3.80	3.03	3.44	n.d	n.d	n.d	3.97	n.d	n.d	n.d	4.16	3.61	3.90	4.11
Element Rations														
Zr/Hf	27.90	24.40	40.60	n.d	n.d	n.d	25.40	n.d	n.d	n.d	36.30	34.60	30.50	31.90
Zr/Y	5.27	5.32	-	3.98	3.64	5.03	5.91	6.95	6.48	5.14	6.07	6.25	5.68	6.81
Zr/Th	8.57	9.22	11.23	n.d	n.d	9.72	9.72	n.d	n.d	n.d	12.82	12.58	12.50	13.92
Cr/Th	2.59	1.23	0.41	n.d	n.d	n.d	1.39	n.d	n.d	n.d	4.76	2.94	4.19	4.34
V/Cr	2.29	4.33	12.80	5.09	10.00	12.00	4.00	1.71	1.62	2.37	1.42	2.02	1.49	1.43
V/Ni	2.42	1.41	2.00	1.65	1.69	1.50	1.63	2.05	2.12	2.08	2.30	3.61	2.17	2.29
Th/U	3.87	3.22	3.55	n.d	n.d	n.d	2.36	n.d	n.d	n.d	4.08	5.37	4.10	4.04
K/Rb	150.0	127.0	98.0	128.0	115.0	111.0	117.0	163.0	164.0	155.0	191.0	169.0	185.0	176.0
Ni/TiO_2	0.0073	0.0119	0.0110	0.0113	0.0119	0.0097	0.0066	0.0065	0.0077	0.0063	0.0059	0.0064	0.0060	0.0240
MgO/TiO_3	8.56	18.19	5.39	17.47	26.21	20.10	16.42	5.14	4.54	5.94	5.00	3.72	5.13	4.40
CaO/TiO_3	20.21	43.91	52.78	46.90	57.42	57.57	38.98	5.37	2.70	6.56	0.87	0.30	1.55	1.56
Na_2O/TiO_2	1.62	1.54	1.02	1.45	1.60	1.55	1.42	1.26	1.18	0.81	1.09	0.65	0.98	1.19
K_2O/Na_2O	6.03	1.15	5.04	3.05	2.27	2.52	2.55	3.33	3.58	6.57	4.01	7.98	4.85	3.47
SiO_2/Al_2O_3	2.65	3.43	2.45	3.82	357	3.88	3.41	3.86	3.99	3.00	3.79	3.30	3.85	3.94

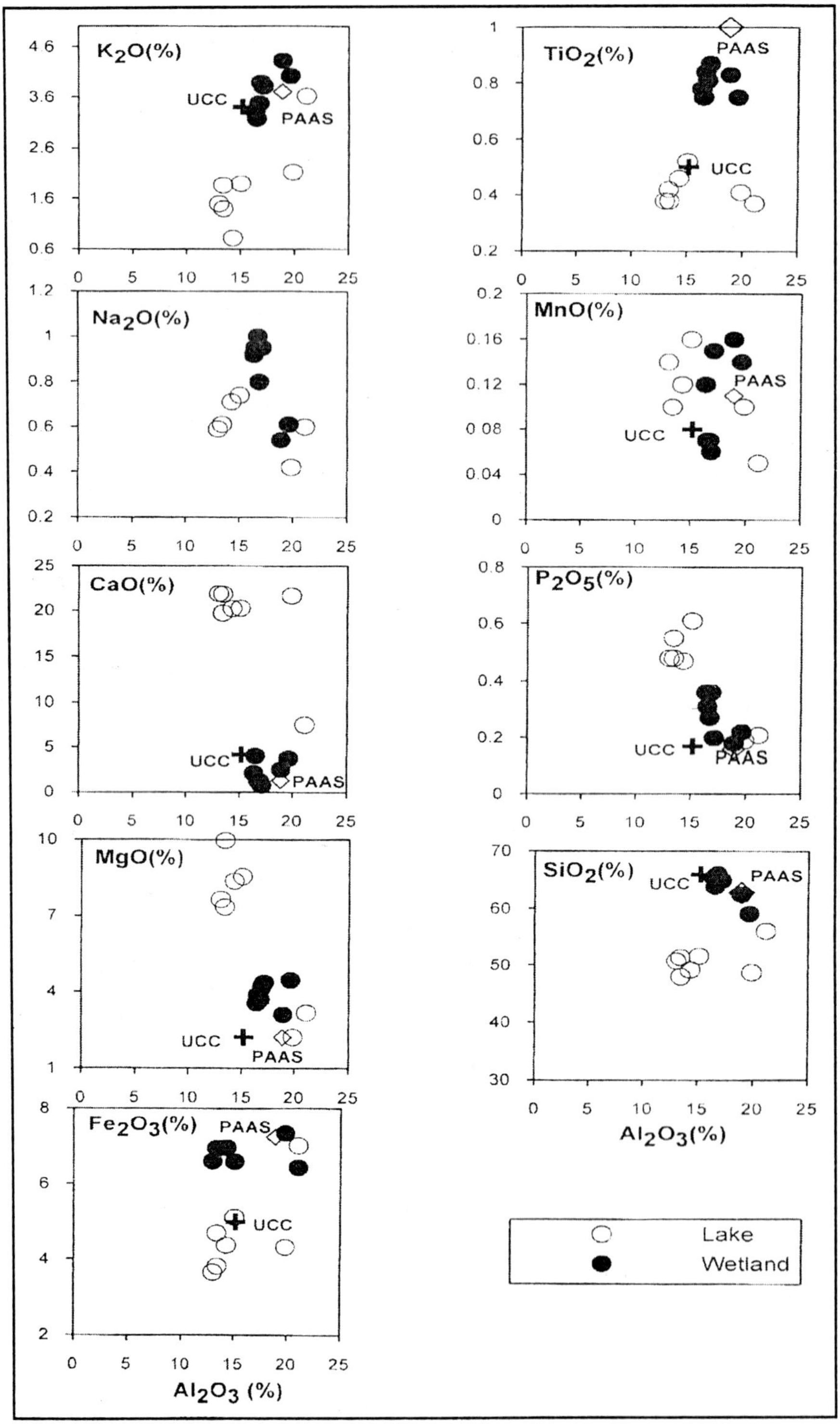

Fig. 7.6. The major oxides are plotted against Al_2O_3. The Renuka sediments show increasing trend with respect to K_2O, Fe_2O_3 TiO_2 and Na_2O from lake to wetland.

lake and is in conformity with the field observations. The Al_2O_3 abundances are used as normalization factor to make possible comparison between different lithologies as it is considered immobile during weathering, diagenesis and metamorphism. The Al-normalized Renuka Lake samples plotted against Al-normalized PAAS (Fig.7.8) show that SiO_2, Al_2O_3, Fe_2O_3, and K_2O are at unity, TiO_2 is very mildly depleted, Na_2O is depleted and MgO and CaO are higher than in PAAS in the wetland samples. In lake samples, SiO_2 and Al_2O_3 are at unity and TiO_2, Fe_2O_3, Na_2O_3 and K_2O are depleted and CaO and MgO are very high. This shows that high CaO in the lake sediments has caused relative depletion in other oxides.

Trace elements concentration in sediments result from the competing influences of provenance, weathering, diagenesis, sediment sorting and the aqueous geochemistry of the individual elements (Rollinson, 1993). Due to remarkably high concentration of these trace elements in clay-rich sediments, most geochemical studies have concentrated on these lithologies (Vital and Stattegger, 2000). Therefore, clays and heavy mineral fractions can be successfully utilized to deduce provenance. The distribution of trace elements in the Renuka Lake and wetland sediments indicate that these elements are in higher concentration in wetland (Clay-rich sediments) than in lake sediments (sand-rich). The Al-normalized values indicate that Sr, Pb, Cu, Zn, and U are higher in lake sediments and Zr, Cr, Nb are depleted and other elements show similar distribution as in wetland and PASS (Fig.7.8). This suggests similar source rocks for the lake as well as for wetland sediments, and the high carbonate content in the lake sediments has contributed high Sr in them as it has chemical affinity for Ca. The Cu and Pb enrichment in the lake sediments are derived from the anthropogenic sources. The variation in distribution of other elements indicate fractionation/grain size sorting of sediments.

The large ion lithophile elements (LILE) Rb, Ba, Th have similar distribution in lake and in wetland sediments as also in PAAS, and Rb and Th have the highest correlation coefficients with Al_2O_3 (r=0.997,0.950 and 0.985, 0.971, respectively). This suggests that their distribution is mainly controlled by phyllosilicates. The high field strength elements Zr, Nb, Ta, Hf, Y, Th, U and REEs which are considered immobile are useful for determination of provenance composition (Taylor and McLennan, 1985). These elements have a similar average concentration in wetland and in PAAS but are depleted in lake sediments indicating Ca dilution effect in case of latter. Forstner and Wittmann (1979) have mentioned that concentration of trace elements in carbonate rocks are greatly reduced compared to shales. A striking depletion can be observed in Ba and Co in lake samples.

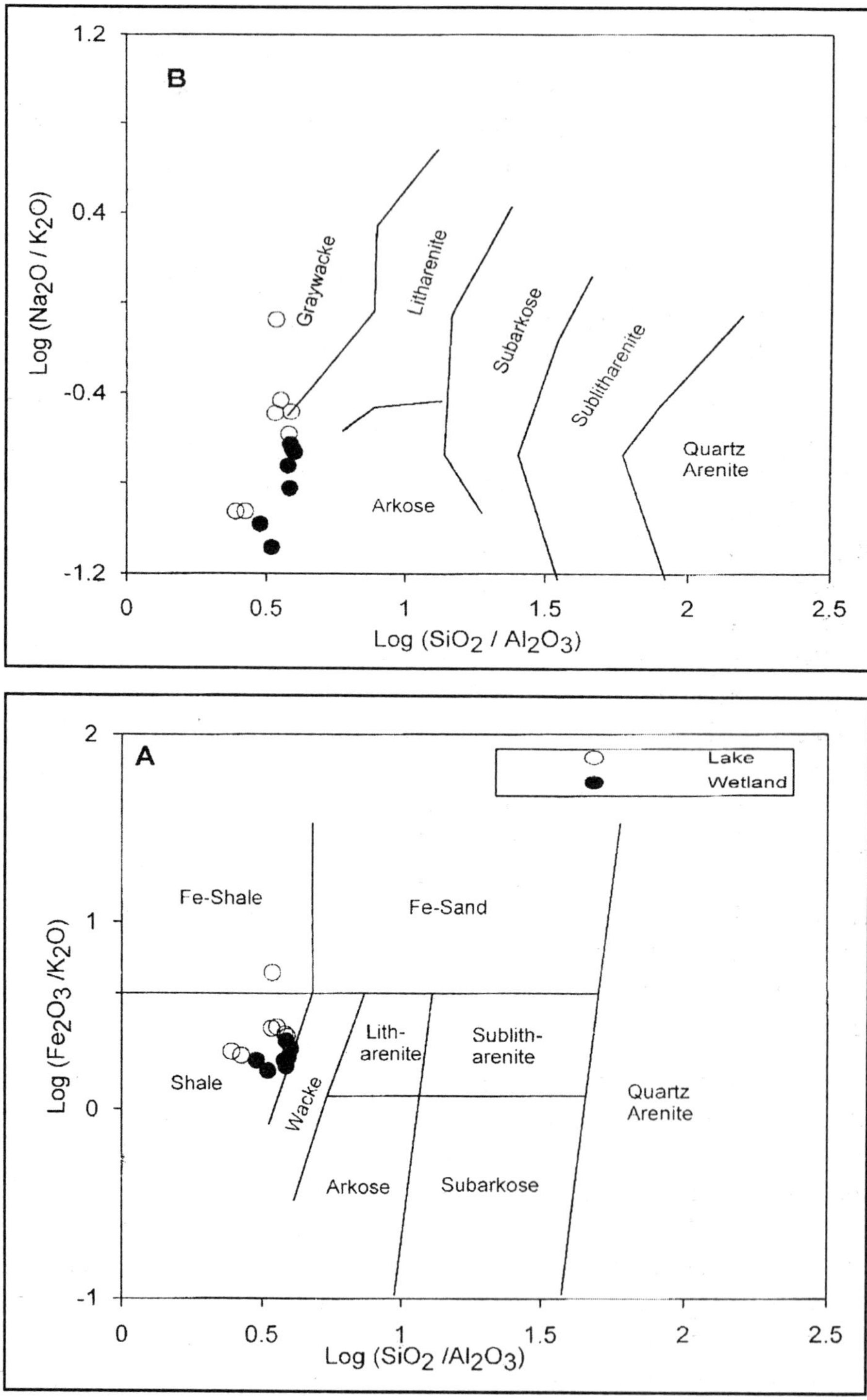

Fig. 7.7 (a,b) The log values of major oxides are plotted in **(a)** Pettijohn *et al.*, (1972) diagram, the Renuka sediments fall in litharenite domain and in **(b)** Herron (1988) diagram it demarcates the lithology of the catachment (Shale, Wacke).

Table 7.4: Rare Earth Elements (REE) data (in ppm) of selected sediment samples of Renuka Lake and wetland.

Sample No.	LAKE				WETLAND			
	GRI	GRII	GRI-II	GRVII	NGRI	NGRIV	NGRV	NGR VI
La	35.4	25.1	28.4	24.9	44.6	46.3	41.6	43.1
Ce	72.8	51.7	58.8	50.4	88.8	94.9	85.1	87.3
Pr	8.08	5.80	6.50	5.75	10-3	10.3	9.51	9.91
Nd	30.5	21.9	24.4	22.1	39.1	38.9	35.9	37.8
Sm	6.24	4.49	5.07	4.48	7.87	8.06	7.62	7.84
Eu	1.18	0.853	0.965	0.920	1.61	1.55	1.43	1.57
Gd	5.48	3.98	4.54	4.36	7.44	7.23	6.95	7.20
Dy	4.81	3.58	3.70	3.81	6.79	6.51	6.36	6.69
Ho	0.976	0.747	0.740	0.769	1.40	1.35	1.3.0	1-35
Er	2.72	2.02	2.13	2.10	3.75	3.80	3.49	3.69
Tm	0.405	0.285	0.294	0.291	0.527	0.560	0.507	0.520
Yb	2.67	1.99	2.14	2.03	3-60	3.78	3.51	3,62
Lu	0.408	0.308	0.333	0.320	0.561	0.592	0.535	0.579
ΣREE	171.67	122.75	138.01	122.23	216.35	223.83	203.81	211.17
La_N/Yb_N	8.96	8.52	8.97	18.29	8.37	8.28	8.01	8,05
Ce/Ce*	1.03	1.03	1.04	1.01	0.99	1.04	1.03	1.01
Eu/Eu*	0.62	0.62	0.62	0.64	0.70	0.62	0.60	0.64
Ce_N/Yb_N	7.07	6.73	7.12	6.43	6,39	6.51	6.28	6,25
Ce_N / Sm_N	2.82	2.78	2.80	2.72	2.72	2.84	2.70	2.69
Gd_N/Yb_N	1.66	1.62	1.72	1.74	1.42	1.55	1.60	1.61
ΣLREE	153.02	108.99	123.17	107.63	190.67	198.46	179.73	185.95
ΣHREE	17.47	12.91	13.88	13.68	24.07	23.82	22.65	23.65
ΣLREE/ ΣHREE	8.76	8.44	8.88	7.87	7.92	8.33	7.93	7.86

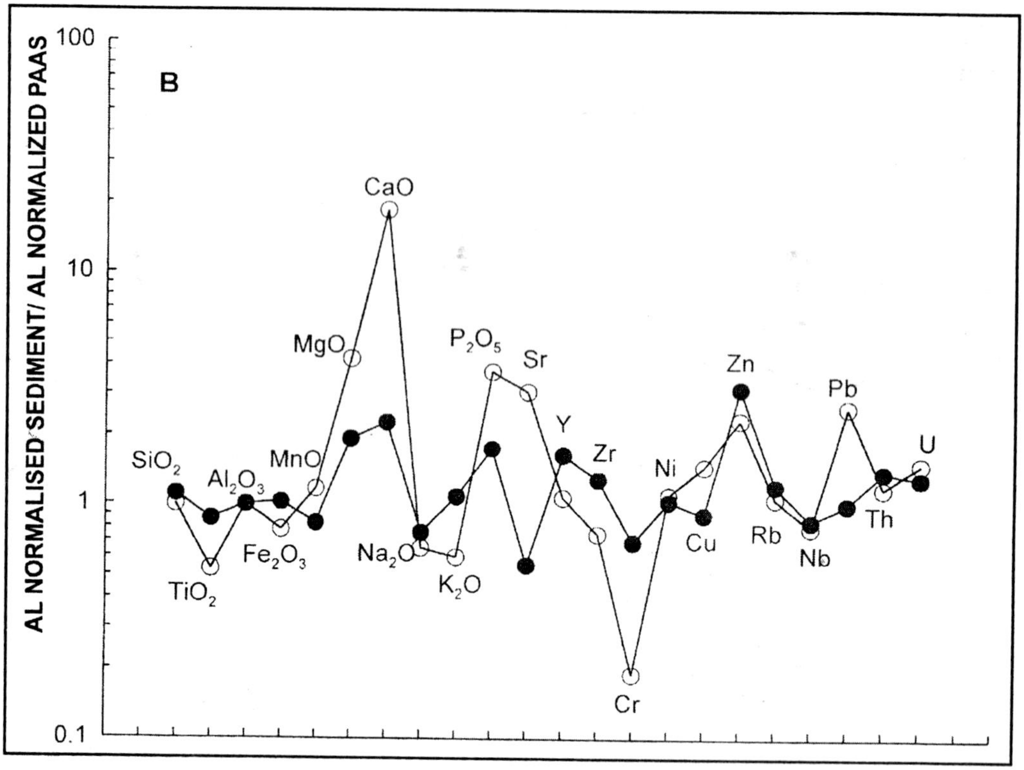

Fig. 7.8. Al-normalized PAAS showing variation in concentration of each elements in lake and wetland sediments.

The wetland sediments are enriched in REEs in comparison to PAAS and UCC whereas lake sediments are mildly depleted (Table 7.4; Fig.7.9 a,b).

However, the chondrite normalized REE pattern of the sediments are essentially similar with abundance decreasing as grain size increases. The steep LREE and almost flat HREE with a steep negative Eu anomaly similar to PAAS and NASC, reflect a cratonic source (7.10a). The individual samples plot show that the wetland sediments cluster while lake sediments are separated suggesting that these are fractionated relative to wetland sediments (Fig.7.10b). The overlapping of HREE profile in the samples may be due to carbonate dilution as these are associated with coarser quartz-rich grains. This is consistent with the major element concentration in the wetland and lake sediments. However, the uniformity of REE distribution in terrigenous sedimentary rocks has been emphasized by McLennan (1989) and is attributed to the efficient mixing of source lithologies during sedimentary processes.

According to Cullers and Stone (1991) the REE content in the clastic sedimentary rocks may be affected by the size sorting and decreases steadily with an increase in the grain size. The wetland sediments show slight LREE

(188.70) and considerable HREE (23.55) enrichment whereas the lake sediments show depletion in both LREE (123.20) and HREE (14.49) in relation to PAAS (Table 7.4). As the LREEs are essentially associated with clay particles and HREEs with heavy minerals, and the lake sediments are considerably Al_2O_3 deficient, consequently significant LREE depletions are not unlikely. As the detritus passes through the wetland during runoff, the sediments are fractionated and those transported into the lake get mildly depleted in HREEs. The HREE enrichment in the wetland sediments, however, suggests particle size sorting during the process of lighter particles (clay) being trapped in the wetland soil. The weathering intensity in the catchment, though higher than PAAS, but is the same in both the cases.

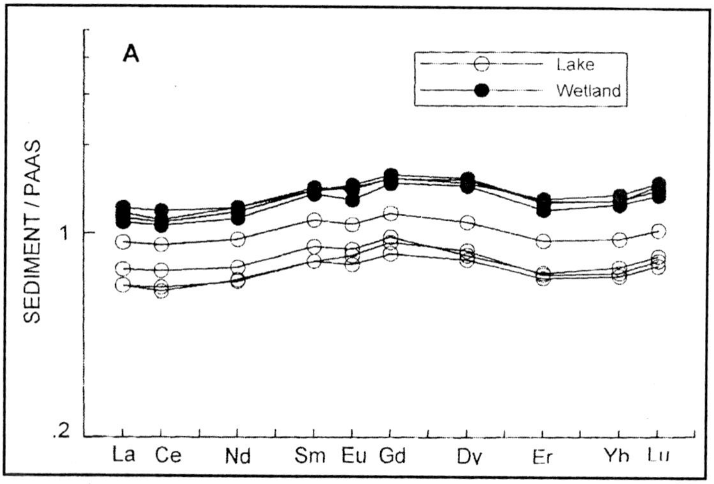

Fig. 7.9 (a). In a PAAS normalized Renuka sediments wetland shows enrichment of REE compared to Lake sediments and clustering of curves.

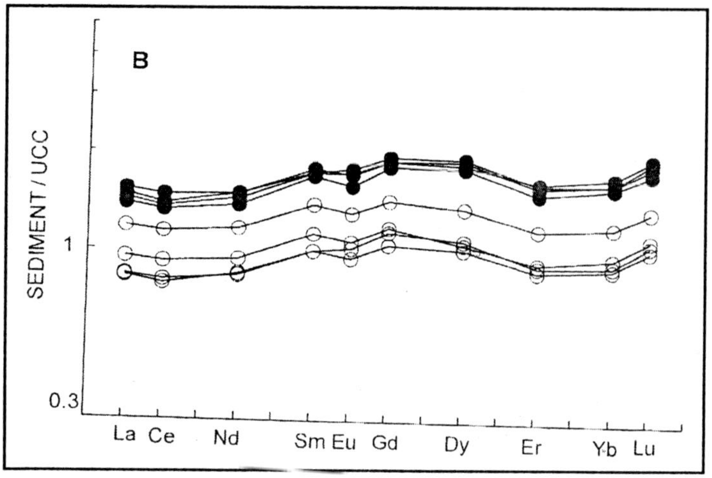

Fig. 7.9 (b). The UCC normalized Renuka sediments show enrichment of REE in wetland compared to lake sediments.

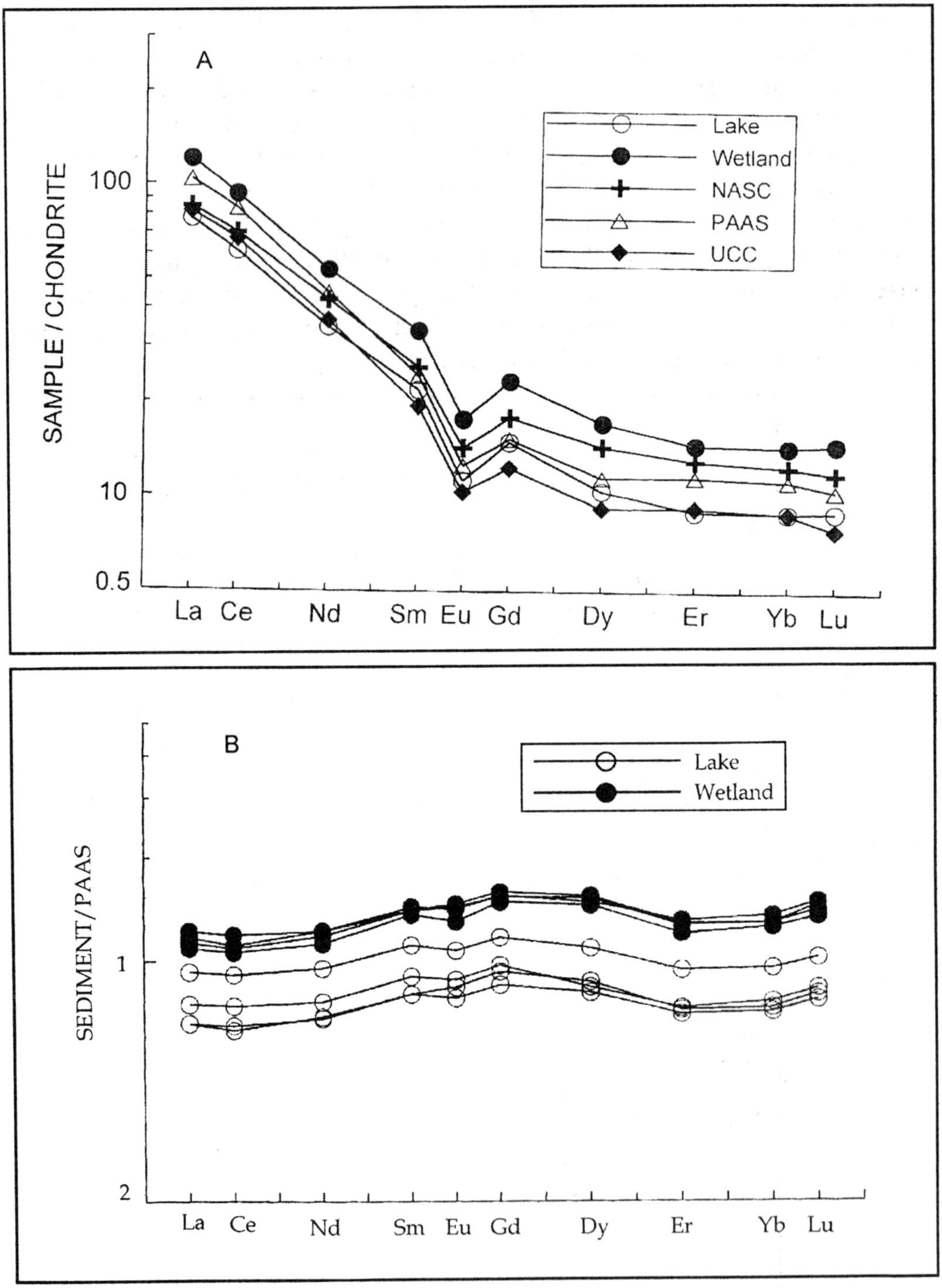

Fig. 7.10 (a,b). Chondrite-normalized REE pattern show (a) LREE enrichment and HREE depletion with negative Eu anomaly, **(b)** the wetland sediments cluster while lake sediments are separated indicating fractionation of sediments and abundance decreases as grain size increases.

The high ΣLREE values for the grab sediments (107.63 to 198.46) and high La/Sm ratio (5.24 to 5.74), along with high total REE values (122.23 to 223.83) are indicative of LREE enrichment, and such results suggests that the sediments are derived from mature cratonic shales and volcanic arcs built on continental crust (Graver and Scott, 1995). Enrichment of REE is usually due to the clay minerals, and as shale-silt are the dominant lithology around Renuka catchment, REE enrichment is quite likely.

The chemical Index of Alteration (CIA) in Renuka Lake and wetland are 78.74% and 72.16% respectively, indicating moderate weathering at source. These values are higher than in shale standard and Post Archaean Australian Shale (PAAS) 69, reflecting recycling or maturity of source sediments. The abundance of clay minerals, illite (72-87%), chlorite (6-13%) and Kaolinite (5-16%) also suggest moderate weathering of the source rocks, hence, consistent with the CIA indices.

Th/U ratio has been used to determine degree of chemical weathering. However, these elements behave differently during weathering as U unlike Th, is chemically mobile (U^{6+}) and therefore, there is a tendency for an elevation of Th/U values above Upper Crustal igneous values of 3.5 to 4.0 (McLennan et al., 1993). The Th/U ratio for lake sediments is 3.25 and that of wetland sediments is 4.39. The higher ratio in wetland sediments is due to higher Th content in the sediment. As the U shows negative correlation with Zr (r=-0.408) in wetland samples and very poor correlation (r=0.113) in lake sediments, its enrichment from Zr is unlikely. Therefore, both these elements are derived from lithogenic sources. The U values in lake and wetland sediments are similar to PAAS, therefore, a moderate chemical weathering in the source area is indicated.

The representation of Renuka sediment data on a binary (K_2O/Na_2O-SiO_2) diagram (Roser and Korsch,1986) show wetland samples falling in PM field and most of the lake samples show a shift towards ACM-PM boundary line (Fig. 7.11a). This shift of lake sediments towards ACM field is due to their low K_2O/Na_2O ratios and low SiO_2 content. The low K_2O is due to sandy nature of sediments and SiO_2 depletion is due to $CaCO_3$ dilution effect. In a discrimination function plot (Roser and Korsch, 1988), wetland sediments overlap and plot on P3 (felsic) and P4 (quartzo-sedimentary provenance) field close to NASC (Fig.7.11b) but the lake sediments plot are scattered with a few samples lying in P3 trending close to P3-P4 boundary and in majority they lie in P4 field. The wetland samples have greater F1 and lesser F2 scores than their counter part lake sediments (Sandy) plotting near the P3-P4 join. Roser and Korsch (1988) have pointed out that although SiO_2 is not used directly in the functions, dilution effect is implicit. These plots, however, have continental signatures and P4 plots reflect cratonic recycled source of sediments deposited in a passive margin tectonic setting.

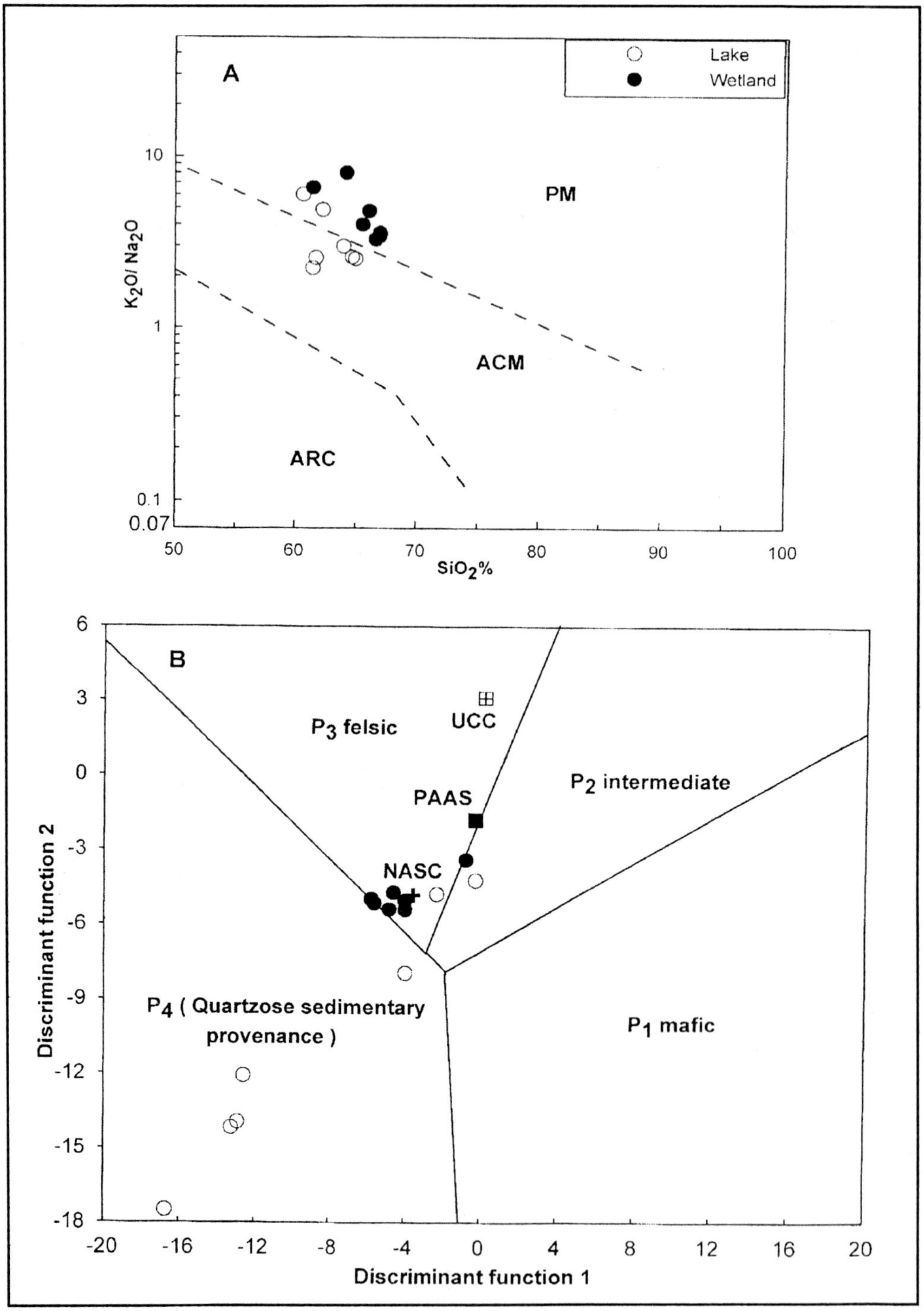

Fig. 7.11 (a,b). In (a) tectonic discrimination diagram (after Roser and Korsch, 1986) wetland sediments plot in PM field and lake sediment in ACM field. While in **(b)** discriminant function plot lake sediments fall in P4 and wetland at the boundary of P3 & P4 close to NASC indicating passive margin setting.

Mansar Lake Sediments

The Mansar Lake in lesser Himalaya, Jammu, is situated on the anticlinal axis of a major anticlinorium of the folded lower Siwalik Formation. The lower Siwalik consists of alternating sandstone, siltstone and shale suggesting that the basin subsided intermittently during the deposition of Siwalik sediments. The lake is non-drainage type and 38.25 m deep. It is at an altitude of 666 m above msl. The sandstone contains megascopic lithic fragments, which are mostly of metamorphic and sedimentary parentage consisting of phyllites, schists, quartzites, slates and chert. Their relatively fair degree of rounding and their soft metamorphic fragments indicate less transport.

The Mansar Lake sediments are high in silica and deficient in Al_2O_3 indicating enrichment of sand fraction (Table 7.5). However, in a plot of major oxides against Al_2O_3, MgO, Fe_2O_3, MnO and P_2O_5 show positive trend whereas CaO, Na_2O and K_2O do not show any trend (Fig. 7.12). This indicates that CaO is derived primarily from the carbonates, and alkalies from feldspars and clay minerals. A comparison of this data with Upper Continental Crust (UCC) and Post Archaen Australian Shale (PAAS) show that SiO_2 is high and TiO_2, Al_2O_3, Fe_2O_3, MnO, P_2O_5 and K_2O are depleted. CaO and Na_2O are higher than PAAS and MgO is similar. This shows that Mansar sediments have not been subjected to intense weathering and recycling. The relatively higher content of CaO may be due to carbonate fraction in the sediments and enrichment of Na_2O may be due to additional sources such as detrital allanite, tourmaline, etc. in the sediments.

The Mansar surface samples identify as litharenite and Arkose and core samples as litharenite on chemical classification diagram of Pettijohn et al., (1972) (Fig.7.13a) and litharenite, wacke and shale on Herron (1988) diagram (Fig 7.13b). Petrographic studies indicate immaturity of lower Siwalik sandstone (Tandon, 1972) and is consistent with the Na_2O/K_2O ratios of detritus which is lower than UCC and higher than PAAS.

Table Contd...

Table 7.5: Major and trace elements in grab sediment samples of Mansar Lake.

Sample No.	Gr-I	Gr-II	Gr-III	Gr-IV	Gr-V	Gr-VI	Gr-VII	Gr-VIII	Gr-IX	Gr-X
SiO_2 (wt%)	75.58	72.61	68.57	70.75	69.16	60.92	66.56	68.64	68.03	54.28
TiO_2	0.36	0.63	0.65	0.66	0.56	0.78	0.68	0.74	0.72	0.73
Al_2O_3	9.36	11.31	11.31	11.58	11.65	15.05	11.96	12.28	11.97	17.92
Fe_2O_3	3.51	2.64	3.35	3.80	3.18	4.32	4.19	4.46	3.65	6.92
MnO	0.05	0.06	0.07	0.06	0.07	0.07	0.06	0.05	0.08	0.08
MgO	2.06	1.71	1.99	1.82	1.80	2.69	2.33	2.39	1.89	2.99
CaO	0.85	1.02	1.59	1.10	1.35	1.76	2.27	0.69	2.45	1.78
Na_2O	1.86	1.61	1.41	1.61	1.52	0.91	1.18	1.30	1.67	3.50
K_2O	2.82	2.42	2.40	2.60	2.50	3.08	3.24	2.38	2.28	1.29
P_2O_5	0.07	0.12	0.15	0.12	0.08	0.15	0.14	0.14	0.15	0.14
LOI	3.51	4.97	8.71	5.55	7.33	9.87	6.69	6.03	6.71	10.77
Total	100.03	99.10	100.20	99.65	99.25	99.60	99.30	99.10	99.60	100.40
CIA	55.00	61.00	61.00	60.00	60.00	70.00	62.00	66.00	60.00	69.00
Rb (ppm)	96.00	96.00	98.00	115.00	107.00	158.00	99.00	108.00	102.00	170.00
Sr	101.00	122.00	126.00	119.00	127.00	90.00	94.00	89.00	146.00	102.00
Y	22.80	30.30	33.20	30.80	26.40	33.60	29.20	29.70	34.00	31.10
Zr	118.00	244.00	342.00	231.00	180.00	214.00	210.00	222.00	332.00	180.00
Pb	19.60	22.80	25.30	27.30	25.20	31.40	22.10	24.40	35.00	93.60
Co	68.00	33.00	25.00	105.00	48.00	26.00	51.00	31.00	12.00	21.00
Ni	25.00	33.00	33.00	36.00	35.00	55.00	46.00	43.00	35.00	122.00
Cr	52.00	52.00	66.00	70.00	52.00	78.00	80.00	60.00	n.d	n.d
Cu	18.00	17.00	19.00	25.00	23.00	41.00	31.00	29.00	23.00	38.00
Zn	36.00	41.00	48.00	61.00	53.00	110.00	63.00	68.00	67.00	123.00
Ga	11.40	12.80	12.70	14.90	13.40	19.40	14.60	16.10	14.80	21.70
Sc	8.00	11.00	10.00	12.00	7.00	15.00	12.00	11.00	9.00	15.00
Nb	9.30	13.20	15.60	13.80	12.20	16.00	12.80	13.80	17.50	14.20
Th	7.90	10.30	12.20	11.00	10.10	13.50	11.00	11.40	13.47	12.20
U	3.00	3.10	3.00	4.50	2.80	5.60	3.10	3.70	3.60	5.70

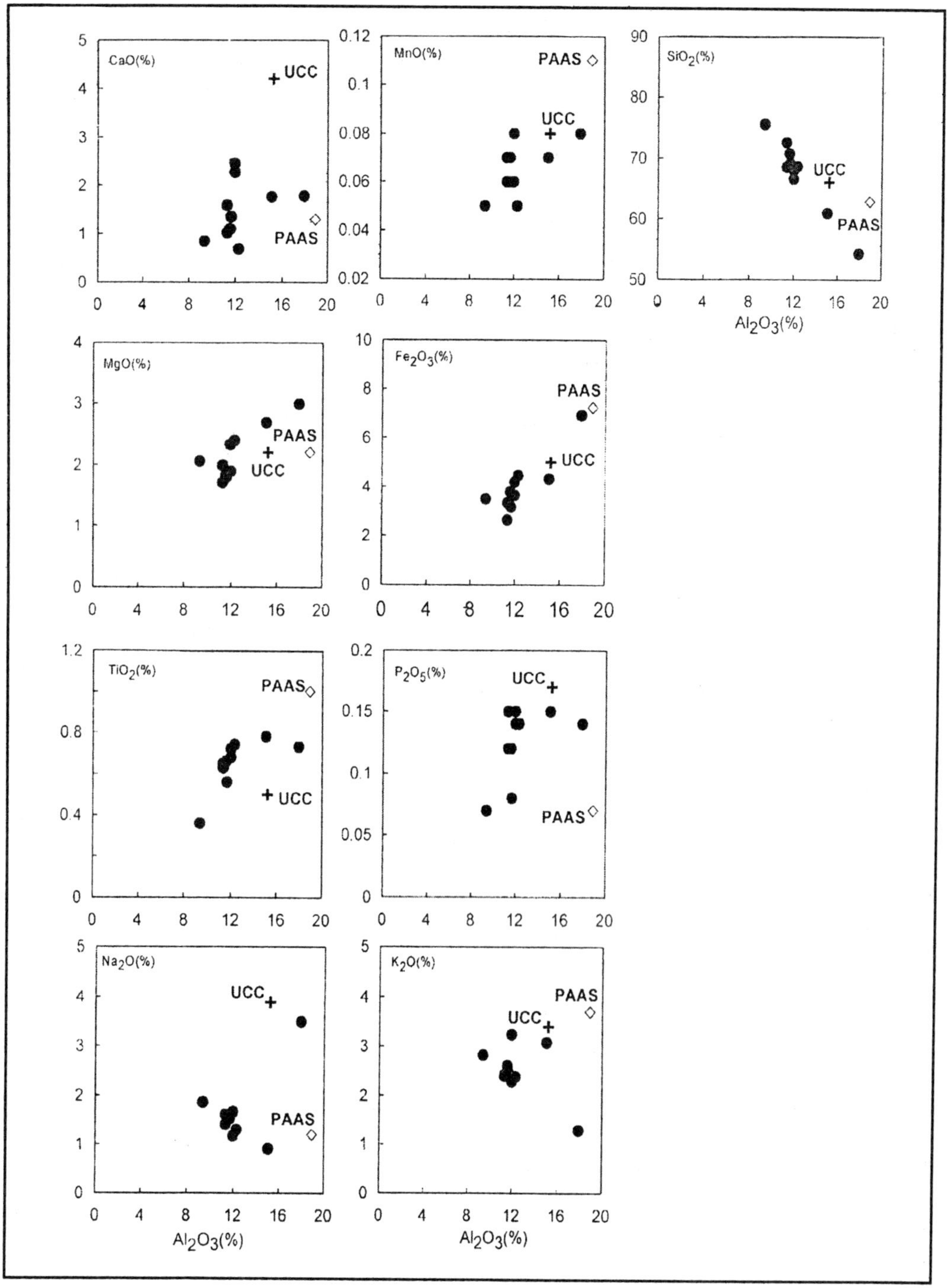

Fig. 7.12. Major oxides plot against Al_2O_3 show a positive correlation with MgO, Fe_2O_3, MnO and P_2O_5, whereas CaO, Na_2O and K_2O do not show any trend. This shows that CaO is derived from carbonate minerals and alkalies from feldspars and clay minerals.

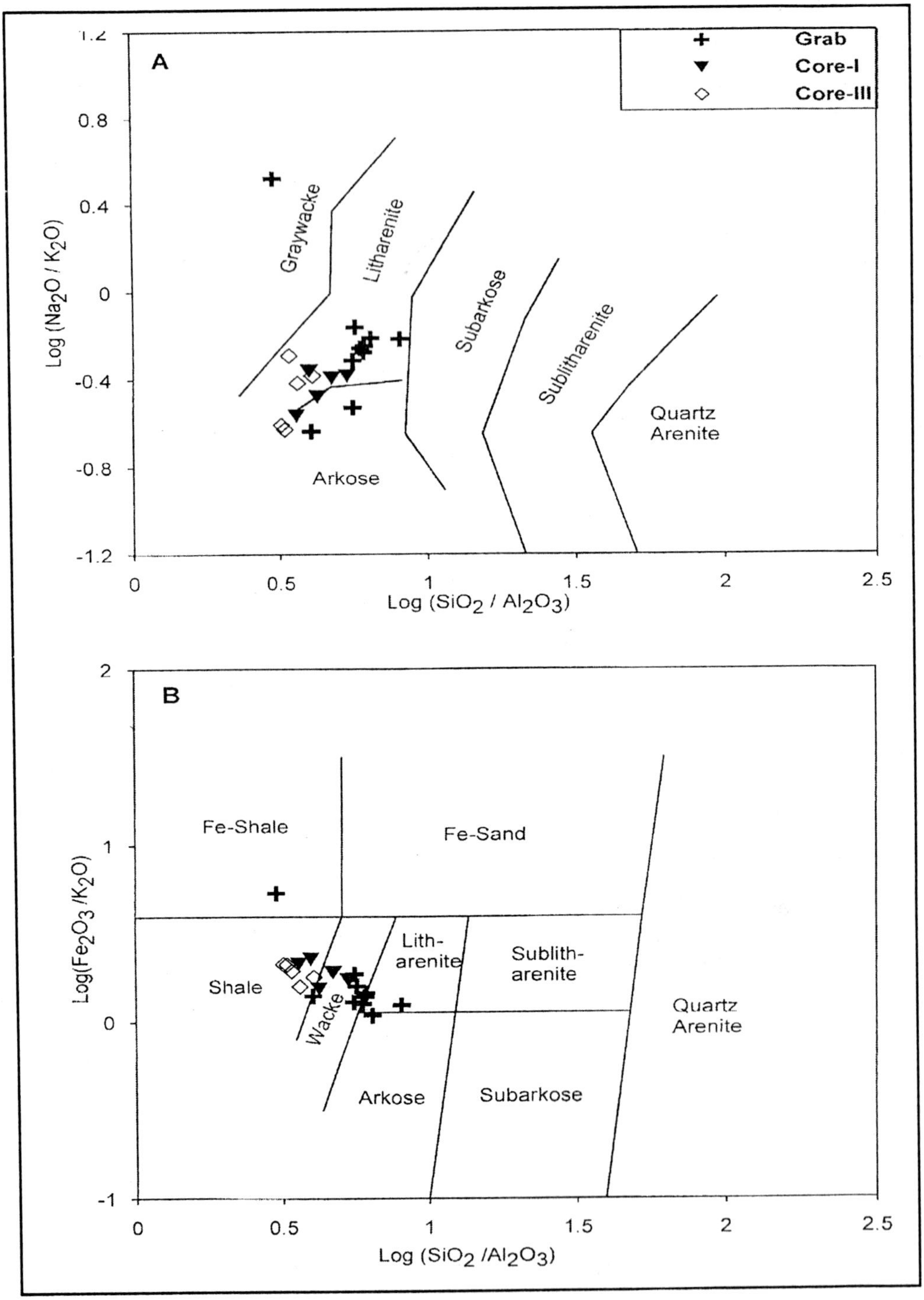

Fig. 7.13 (a,b). On geochemical classification diagram of Pettijohn et al., (1972) and Herron, (1988) the Mansar Lake sediments fall in litharenite and Arkose and core samples in litharenite field in Pettijohn et al. (1972) diagram and in litharenite, wacke and shale fields in Herron (1988) diagram.

Mansar sediments show enrichment in some elements such as Pb (1.63), ratios <1. On average Mansar sediments have similar Zr (227 ppm) and Y (30 ppm) contents as PAAS (210 and 27 ppm, respectively) (Table 7.5). Mineralogical studies have shown that coarse fractions are most significant in the lake, mainly due to their high quartz and feldspar content which are resistant to weathering and are poor in transition elements. The high Zr contents may indicate recycled or fractionated sediments (Basu et al., 1982; McLennan et al., 1993) as evident by their enrichment in zircon. Ni, Cu, and Cr correlate positively with Al_2O_3, indicating preferential incorporation into clay fractions and are presumably lithogenic. Generally, transition elements are adsorbed on fine-grained sediments and decrease in concentration with increasing grain size. Chondrite-normalized REE plot of Mansar Lake sediments show similar REE patterns, LREE enrichment almost flat HREE and negative Eu anomaly (Fig. 7.14a,b) (Table 7.6).

The patterns are in general similar to the chondrite normalized PAAS and UCC indicating cratonic source (McLennan and Taylor, 1991). The persistence of uniform REE patterns of individual samples also indicates uniform source and/or homogenization of the sediments (McLennan, 1989). This supports the quantitative analysis data of Siwalik Group samples plotted on a triangular (QmFLt) graph, demonstrating that sediments were derived primarily from the cratonic interior domain (Fig.7.15). Deformed biotite flakes, staurolite and kyanite as heavy minerals in the detritus, which are characteristic of medium to high-grade metamorphic rocks, supports metasedimentary source rock for the detritus.

The large ion lithophile elements Rb, Sr, Th and U, except Sr, show significant correlations (r=0.94, 0.85 and 0.60, respectively) with Al suggesting that the distribution of these elements are mainly controlled by phyllosilicates and are lithogenic. The high concentration of U (4.4 ppm) in the sediments may be derived from the source rock and/or heavy minerals such as zircon and allanite (Vital and Stattegger, 2000).

The Chemical Index of Alteration (CIA) varies from 66.48 to 78.43 in grab samples whereas in cores it varies between 77.92 to 85.45 and 76.60 to 86.46. The lake sediments therefore show moderate chemical weathering at source area. The higher values in core samples are due to greater abundance of Al_2O_3 and lesser SiO_2, Na_2O and K_2O. The clay mineral contents in order of decreasing abundance of illite, chlorite and kaolinite indicate a cold humid climate in the source area and moderate chemical weathering.

Previous studies have shown that REEs, Th, Sc and high field strength (HFS) elements are specially useful for monitering source-area composition. A higher Zr/Th ratio close to 1 is typical of continental crust that is enriched in incompatible elements and therefore indicates a cratonic source for the detritus.

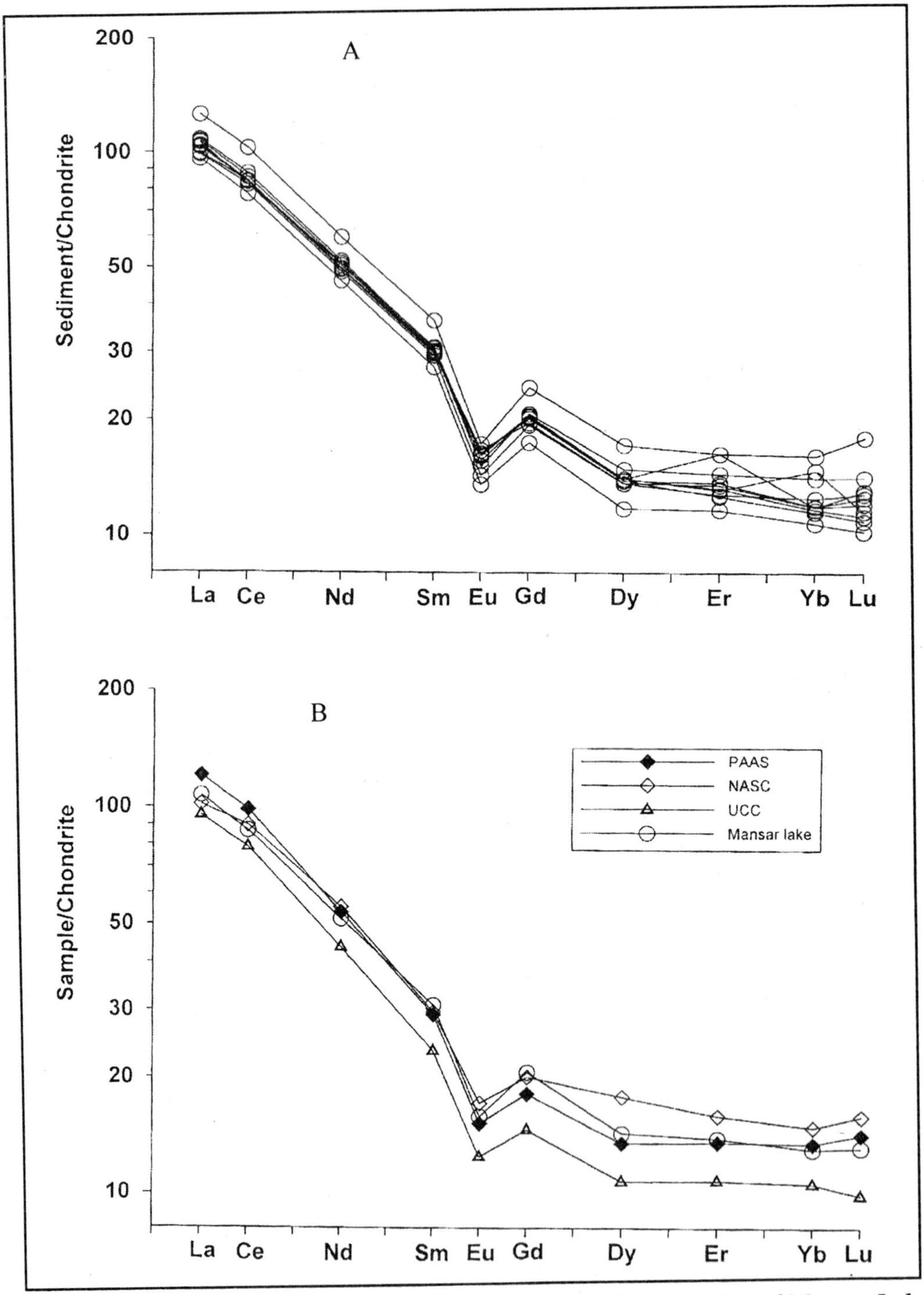

Fig. 7.14 (a,b). Chondrite normalized REE plot of individual samples of Mansar Lake show steep LREE and gentle HREE pattern with steep Eu anomaly and clustering of plots in LREE.(b) Chondrite normalized pattern show similarity with PAAS, NASC, and UCC indicting cratonic source of sediments.

Table 7.6: Rare earth elements (REE) in Mansar grab samples.

La	34.10	n.d	40.22	33.76	30.25	n.d	31.10	31.42	33.48	n.d
Ce	71.60	n.d	84.34	69.85	62.93	n.d	66.80	68.18	66.85	n.d
Nd	30.98	n.d	36.44	30.53	27.44	n.d	29.00	30.08	30.50	n.d
Sm	5.88	n.d	7.11	5.92	5.24	n.d	5.60	5.78	5.81	n.d
Eu	1.09	n.d	1.33	1.16	0.98	n.d	1.00	1.20	1.13	n.d
Gd	5.35	n.d	6.40	5.30	4.51	n.d	5.00	5.06	5.28	n.d
Dy	4.83	n.d	5.88	4.57	3.82	n.d	4.40	4.45	4.51	n.d
Er	3.08	n.d	3.83	2.80	2.48	n.d	2.70	2.88	2.92	n.d
Yb	2.93	n.d	3.64	3.07	2.23	n.d	2.60	2.47	2.48	n.d
Lu	0.43	n.d	0.61	0.35	0.31	n.d	0.30	0.40	0.38	n.d
REE	160.20	n.d	189.80	157.31	140.20	n.d	148.70	151.92	123.40	n.d
La_N/Yb_N	7.68	n.d	7.30	7.26	8.96	n.d	7.90	8.40	8.92	n.d
Ce/Ce*	1.18	n.d	1.18	1.16	1.17	n.d	1.10	1.18	1.12	n.d
Eu/Eu*	0.60	n.d	0.61	0.64	0.62	n.d	0.60	0.68	0.63	n.d
Ce_N/Yb_N	6.25	n.d	5.93	5.82	7.23	n.d	6.50	7.06	6.90	n.d
Ce_N/Sm_N	2.87	n.d	2.80	2.79	2.84	n.d	2.80	2.79	2.72	n.d
Gd_N/Yb_N	1.47	n.d	1.41	1.39	1.62	n.d	1.50	1.64	1.71	n.d
LREE	142.60	n.d	168.11	140.06	125.86	n.d	132.50	135.46	136.64	n.d
HREE	16.62	n.d	20.36	16.09	13.50	n.d	15.00	15.26	15.57	n.d
LREE/HREE	8.58	n.d	8.26	8.70	9.43	n.d	8.70	8.88	8.77	n.d
K/Rb	243.95	209.37	203.06	187.83	193.46	162.02	271.72	182.41	185.29	62.94
Th/U	2.63	3.32	4.07	2.44	3.61	2.41	3.55	3.08	3.81	2.20
Th/Sc	0.99	0.94	1.22	0.91	1.44	0.90	0.92	1.04	1.52	0.80
Zr/Sc	14.75	22.18	34.20	19.25	25.71	14.27	17.50	20.18	36.89	12.00
La/Th	4.32	n.d	3.30	3.07	3.00	n.d	2.86	2.93	n.d	n.d
La/Sc	4.26	n.d	4.02	2.81	4.32	n.d	2.62	3.04	n.d	n.d
Sc/Th	1.01	1.10	0.82	1.09	0.69	1.10	1.09	0.96	0.66	1.20
Cr/Ni	2.08	1.58	2.00	1.94	1.94	1.42	1.74	1.39	n.d	n.d
Cr/Th	6.82	5.05	5.41	6.36	5.15	5.78	7.27	5.26	n.d	n.d
Zr/Th	14.94	23.69	28.03	21.00	17.82	15.85	19.09	19.47	24.23	14.75

n.d=not determined

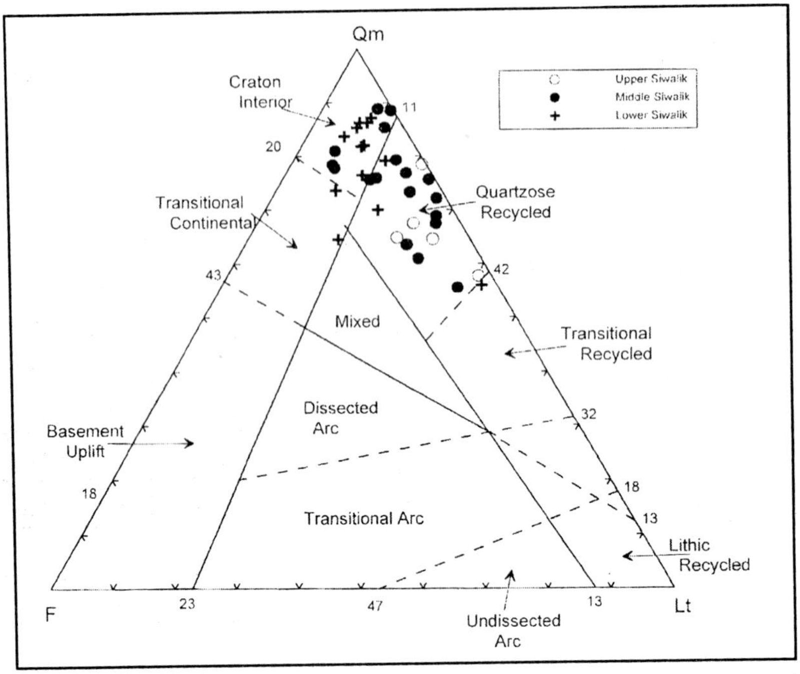

Fig. 7.15. In a QmFLt plot showing the petrographical composition of the Siwalik Group in relation to provenance fields, the lower Siwalik samples fall in the cratonic interior domain.

The plot of grab samples in La-Th-Sc compositional space show mixed sedimentary field and samples plot close to UCC and PAAS, indicating that the sediments were derived from a source similar to mixed sedimentary-metasedimentary provenance and not from a felsic igneous source (Cullers, 1994) (Fig.7.16). The plots are closer to the La apex and to the UCC and PAAS, indicating that despite weathering and sediment transport, La, Th and Sc remained immobile.

According to McLennan et al. (1993), the Th/Sc ratio is a sensitive indicator of the bulk composition of the provenance and the Zr/Sc ratio is useful index of zircon enrichment. In Fig. 7.17 most of the points cluster between UCC and PAAS except for three sand-rich samples having higher Zr/Sc ratios, indicating sedimentary sorting and mild recycling.

The Mansar Lake sediments show a weak positive correlation with SiO_2 and a poor negative correlation with Al_2O_3 between light and heavy REEs, suggesting that the REEs are contributed by both heavy minerals (particularly zircon and sphene) and fine clay fractions. The very low negative correlation of REEs with major oxides may suggest uncommon sources, i.e. heavy minerals as well as clay minerals as discussed above. Therefore, Zr does not dominantly control the distribution of REEs in sediments, but there is mixed contribution of REEs in Mansar detritus.

The plot of Mansar samples on tectonic discrimination diagrams show that majority of samples plot within the P4 quartzose sedimentary provenance field, which is similar to a PM-derived source, suggesting that they were derived from a cratonic interior or a recycled orogenic terrain (Fig.7.18). However, as discussed above, cratonic interior seems to be the most probable source which is supported by the QmFLt plot of lower Siwalik rocks.

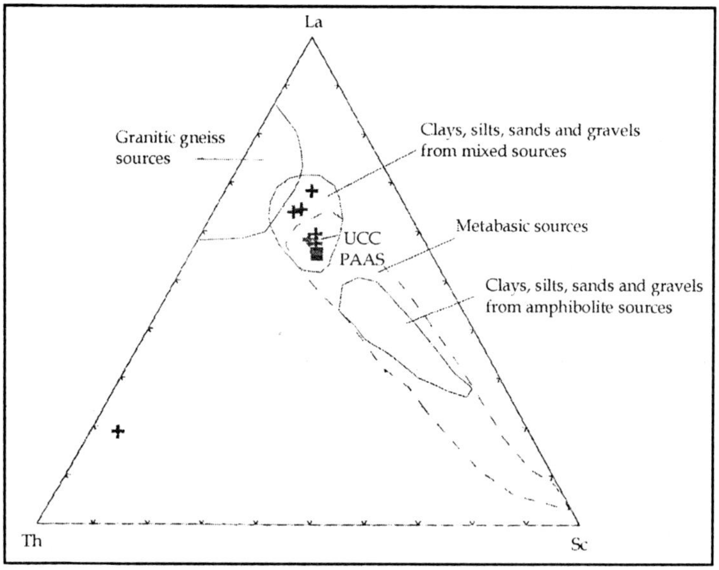

Fig. 7.16. In a La-Th-Sc ternary plot, Mansar sediments plot in a region of mixed sedimentary source field.

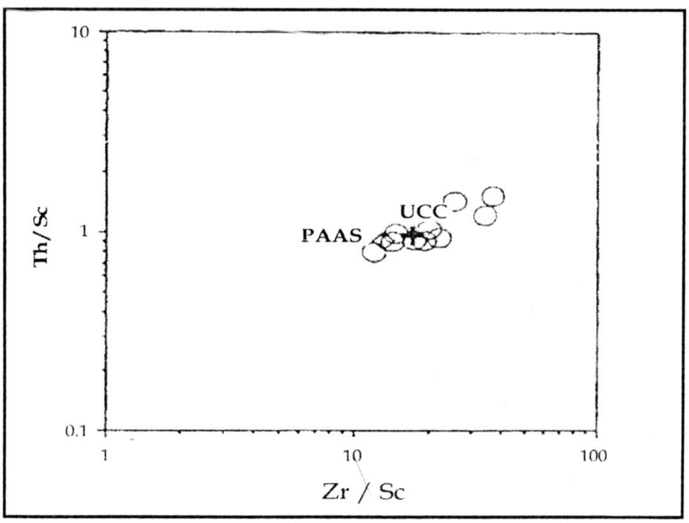

Fig. 7.17. Plot of Th/Sc vs Zr/Sc ratios of Mansar sediments show clustering between UCC and PAAS except three sand rich samples plotting away having higher Zr/Sc ratios which show sedimentary sorting.

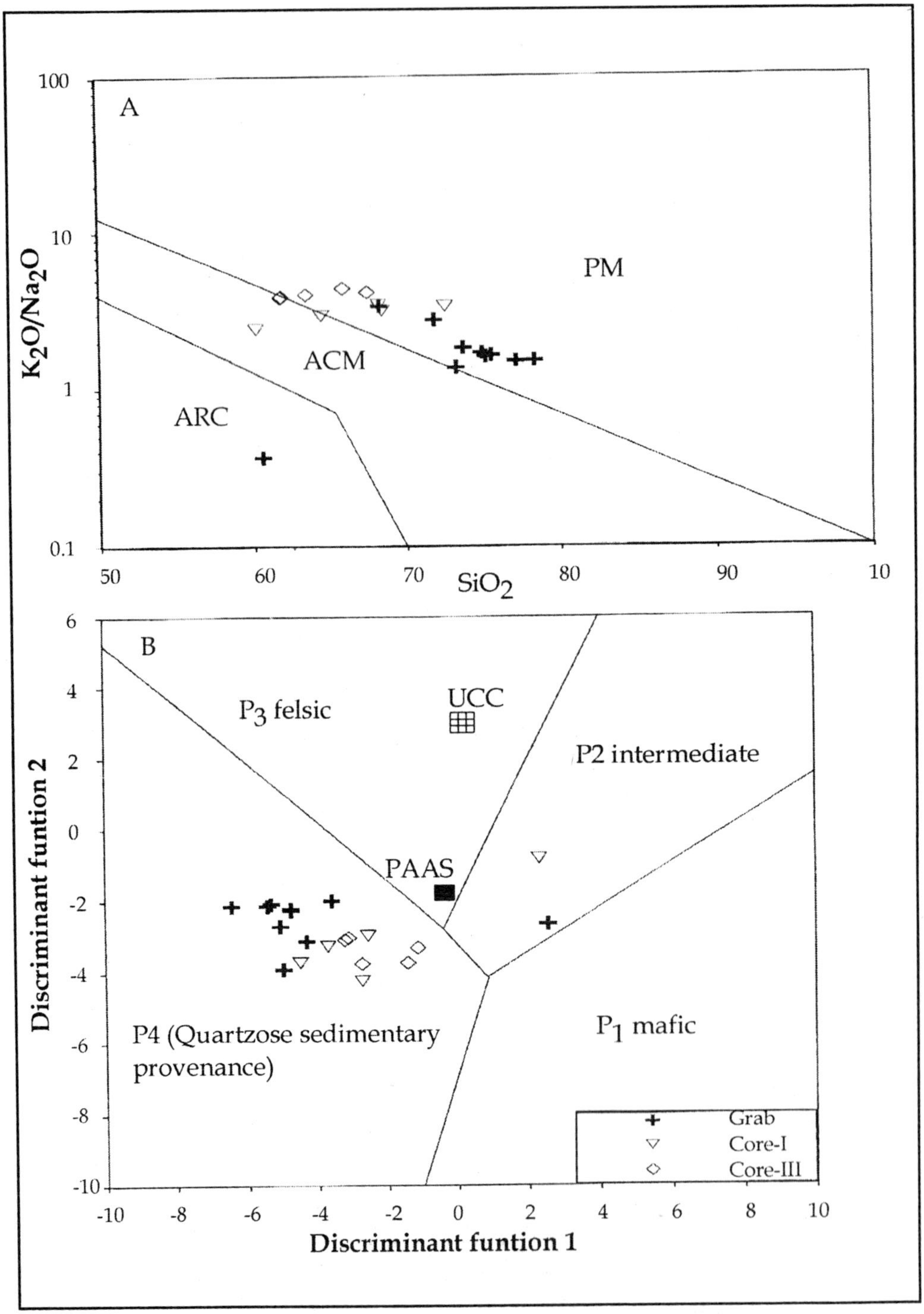

Fig. 7.18 **(a)** In a tectonic discrimination diagram Mansar sediments plot in passive margin field, **(b)** in discrimination function diagram the samples lie in P4 field (quartzose sedimentary provenance) domain.

Dankar, Gete and Thinam Lakes of Spiti

The Spiti valley is unique in comprising Tethyan Sediments and preserving a complete record of sedimentation from Cambrian to Cretaceous. The catchment of Dankar Lake comprises of the Lilang Group of rocks (Triassic) consisting of black limestone with shale intercalations, dolomite and sandstone. These are fossiliferous. Thinam and Gete are on the Spiti Formation (Jurassic), and the latter is at the contact of Lilang Group, comprising black shales, fine-grained sandstone and siltstone. The lustrous black shales contain abundant calcareous concretions enclosing fossils. The nodules are phosphatic and calcareous. The black shale sporadically shows gypsum scales. There are a number of thin limestone bands and calcareous sandstone within the black shale.

The major element chemistry of Dankar and Gete lakes has been recalculated to 100% volatile free on anhydrous basis as the sediments are rich in carbonate content (Das & Dhiman, 2003). The Data of Dankar sediment show wide variation in SiO_2 and in CaO but variation is not so marked in Al_2O_3 and other oxides (Table 7.7). It is observed that as the CaO content increases, SiO_2 decreases. This shows dilution effect of carbonate on SiO_2. The other oxides redistributed accordingly. However, this data compares well with average shale except that CaO and MgO are much higher in lake sediments. A comparison of data with those of Post Archaean Australian shale (PAAS) and North American shale composite (NASC) show that SiO_2, Na_2O and K_2O are depleted and the former show pronounce depletion. Similarly, CaO is very much in excess. Al_2O_3 is comparable with them. This again shows SiO_2 depletion due to carbonate dilution. A strong negative correlation of CaO and MgO with SiO_2 (r = - 0.98 and –0.88, respectively) and Al_2O_3 (r = -0.48 and -0.47, respectively) suggest carbonate source rock for CaO and MgO. The Al_2O_3 abundances have been used as the normalization factor to make possible comparison between different lithologies, as it is immobile during weathering, diagenesis and metamorphism (Cardenas *et al.*, 1996). The Al-normalized average sediments show SiO_2, Al_2O_3 and P_2O_5 close to unity, whereas TiO_2, Fe_2O_3, MnO, Na_2O and K_2O are depleted and CaO, MgO are enriched as compared to PAAS (Fig.7.19). This shows strong weathering at source, removing mobile elements and predominance of carbonate lithology, contributing CaO and MgO to the sediments. The decrease in K_2O is attributed to depletion in K-feldspar and consequent maturity of detritus. Such a geochemical character has been ascribed to indicate mineralogical maturity and recycling of the sediments (Bhatia, 1983).

Table 7.7: LOI Free Major Oxides Composition of Grab Sediments of Dankar and Gete Lakes

Oxides	D-1	D-2	D-3	D-4	D-5	D-6	D-9	Average	G-1	G-2	Average
SiO_2	59.27	39.24	72.19	56.85	55.85	56.47	43.64	54.78	50.25	47.33	48.79
Al_2O_3	19.46	18.06	15.20	14.93	13.95	15.54	17.84	16.43	17.51	15.97	16.74
TiO_2	0.79	0.33	0.84	0.55	0.54	0.55	0.42	0.57	0.51	0.38	0.45
Fe_2O_3	5.18	2.49	4.44	4.02	3.59	4.38	2.79	3.86	4.44	3.77	4.10
MgO	1.31	4.67	1.17	2.72	2.78	2.80	4.67	2.87	3.39	2.91	3.15
CaO	10.35	33.01	1.65	18.25	20.69	17.33	28.82	18.59	21.27	27.91	24.6
Na2O	0.53	0.59	0.56	089	0.95	0.91	0.65	0.78	0.74	0.88	0.81
K_2O	2.94	0.74	3.75	1.56	1.33	1.76	0.98	1.87	1.64	0.69	1.16
MnO	0.02	0.04	0.03	0.06	0.05	0.06	0.05	0.04	0.05	0.04	0.04
P_2O_5	0.16	0.15	0.15	0.16	0.16	0.16	0.15	0.15	0.18	0.08	0.13
Total	100.01	99.31	99.98	99.99	99.99	99.96	100.01	99.94	99.98	99.96	99.97
K_2O/Na_2O	5.55	1.25	6.70	1.75	1.40	1.93	1.50	3.44	2.22	0.78	1.50
Fe_2O_3+MgO	6.49	7.16	5.61	6.74	6.47	7.18	7.46	6.73	4.49	3.81	4.15
Al_2O_3/SiO_2	0.33	0.46	0.21	0.26	0.25	0.27	0.41	0.31	0.35	0.34	0.34
Al_2O_3/Na_2O+CaO	1.79	0.54	6.88	0.78	0.64	0.85	0.60	1.73	0.79	0.55	0.67
Na_2O + CaO	10.88	33.60	2.21	19.14	21.64	18.24	29.47	19.31	22.01	28.8	25.40
TRACE ELEMENTS (PPM)											
Ba	202	<5	225	32	10	26	60	64.67	10	<5	7.5
Cr	828	7	66	40	22	21	27	30.50	21	13	17.0

Contd....

Table 7.7 Contd....

V	58	38	77	54	49	57	49	54.00	56	43	49.5
Ni	245	33	35	42	37	39	36	37.00	43	40	41.5
Cu	19	19	17	26	25	29	20	22.70	29	25	27.0
Zn	74	63	80	71	66	80	68	71.30	85	62	73.5
Pb	33.7	25.4	33.3	32	28.6	34	25.2	29.80	34	23.3	28.65
Th	14.5	10.2	16.0	12.2	12.3	12.0	11.2	12.32	11.3	10.7	11.00
Rb	100	70	113	79	73	94	79	84.74	103	69	86.00
Sr	188	1,957	122	216	251	250	2,235	838.50	600	841	720.5
Y	22.3	-	26.3	18.1	16.0	18	-	14.73	12.4	11.2	11.8
Zr	403	248	421	371	367	300	262	328.17	236	201	218.5
Nb	18.3	10.0	21.5	14.2	14.3	14.0	11.8	14.30	12.4	11.2	11.8

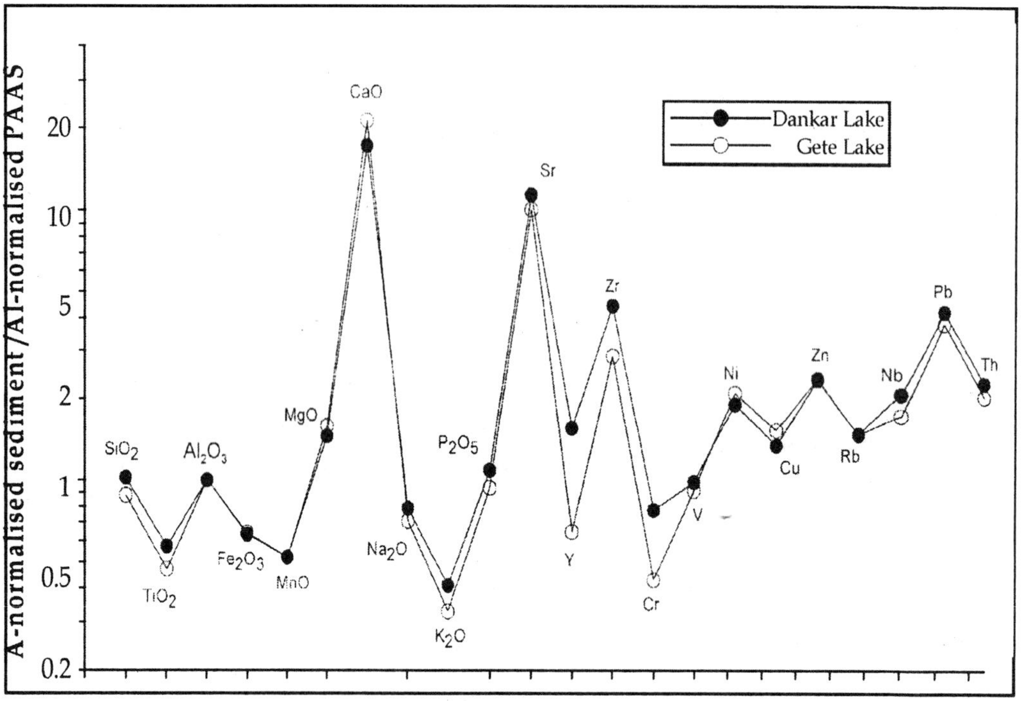

Fig. 7.19. In Al-normalized sediments against PAAS Dankar and Gete sediments delineates different lithologies as alumina is considered immobile.

In a PAAS normalized multi-element plot of Dankar and Gete sediments almost all major and trace elements show depletion except for MgO, CaO, P_2O_5, Sr, Zr, Pb, and Fe_2O_3 in case of Gete, in relation to PAAS. This shows stronger weathering in the source area and very high CaO and Sr suggest carbonate source rock for these elements which also caused dilution of other elements specially SiO_2. Similarly, strong positive correlation of Th and Ba with K_2O (r = 0.98 and 0.94, respectively) may be attributed to their association with shaly rocks. However, low content of Ba and Rb in both the lakes in relation to PAAS, may be due to dilution effect of $CaCO_3$.

The high field strength elements (HFSE) Zr, Hf, Y and Nb are immobile in behaviour and are considered to reflect provenance composition (Taylor and McLennan, 1985: McLennan *et al.*, 1983, 1990; Cullers, 1994). The Zr content in Dankar sediments is much higher (328 ppm) as compared to PAAS (210 ppm). Its high positive correlation with SiO_2 (r= 0.87), weak correlation with Al_2O_3 (r = 0.30), and strong positive correlation of Y with Si (r = 0.91) and with Zr (r = 0.91) indicates its association with minerals preferentially concentrated in quartz-rich rocks. The Nb content (14.3 and 11.8 ppm, respectively) in Dankar and Gete sediments are slightly lower than that of

PAAS (19 ppm) but comparable with the NASC 13 ppm. It is positively correlated with Y (r = 0.88) and Zr (r = 0.90) indicating that Y and Nb are contained in Zr. In Al-normalized sediments, Zr, Y and Nb show higher concentration than PAAS, suggesting their abundance in sandstone lithology at the source, and mineral sorting concentrating zircon which commonly occurs as a detrital phase.

The transition trace elements Cr, Ni, V and Cu are also depleted in comparison to PAAS, but are comparable to UCC indicating stronger weathering at the source. The CIA value in lake sediments ranges from~72 to 87, with an average of ~ 79 for the Dankar Lake and 80 to 81 in Lake Gete. This value for PAAS is reported to be 69. Therefore, moderate to strong weathering conditions in source area of both the lakes are reflected. These samples on the K_2O-Fe_2O_3-Al_2O_3 diagram (Fig. 7.20) plot in the residual clay

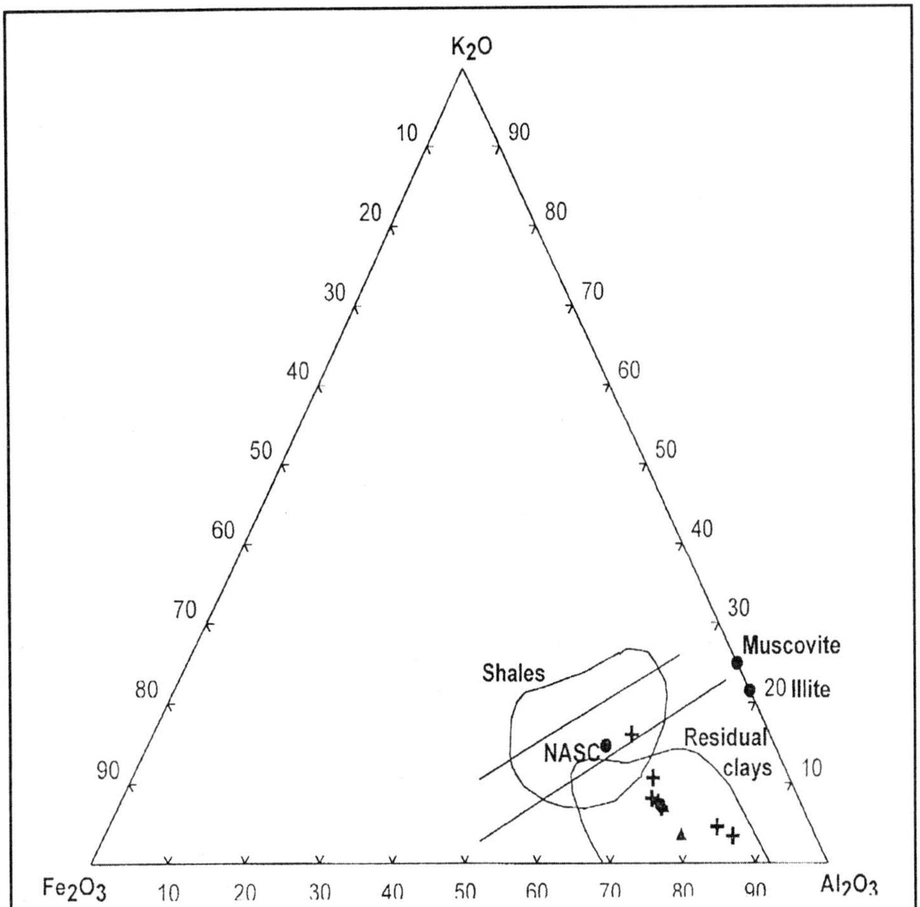

Fig. 7.20 K_2O-Fe_2O_3 Al_2O_3 triangular plot of lake sediments which falls in the residual clay field indicating intense weathering.

field and thus corroborate the high CIA values of the lake sediments. It suggests that these shales are not derived from the felsic-rich source area, as the shales with the highest CIA values have the lowest K_2O/ Na_2O ratio. This indicates that these elements are hosted by clay minerals and not by detrital feldspars (Cox *et al.*, 1995). The depletion of Na is attributed to low abundance of feldspars in source rocks, and loss during transportation and weathering.

The major element chemistry gives clues as to the provenance type as well as weathering conditions, both of which are controlled by the tectonic setting of the basin (Bhatia, 1983). The most discriminating parameters i.e. lower Fe_2O_3 + MgO (av. 6.73 and 4.15, respectively) in Dankar and Gete sediments, and higher K_2O/Na_2O ratios (av. 3.44 and 1.50, respectively) represent chemical characteristics of passive margin setting. By contrast, the higher Al_2O_3/SiO_2 ratio 0.31 and 0.34 respectively, in the two lakes and low $Al_2O_3/(CaO + Na_2O)$ (av. 1.73 and 0.67, respectively) are due to low SiO_2 and high CaO in the terrigenous detritus of these lakes. Such a departure is reported in modern deep sea sandstones of the trailing margins where a low Al_2O_3/ (CaO + Na_2O) ratio has been considered due to presence of biogenic carbonates (Bhatia, 1983). The bivariate plot K_2O/Na_2O vs SiO_2 discriminates tectonic fields of the basin. Lake Sediments plot in both passive and active margin fields (Fig. 7.21). The silica-rich samples (sandy) fall in PM field while

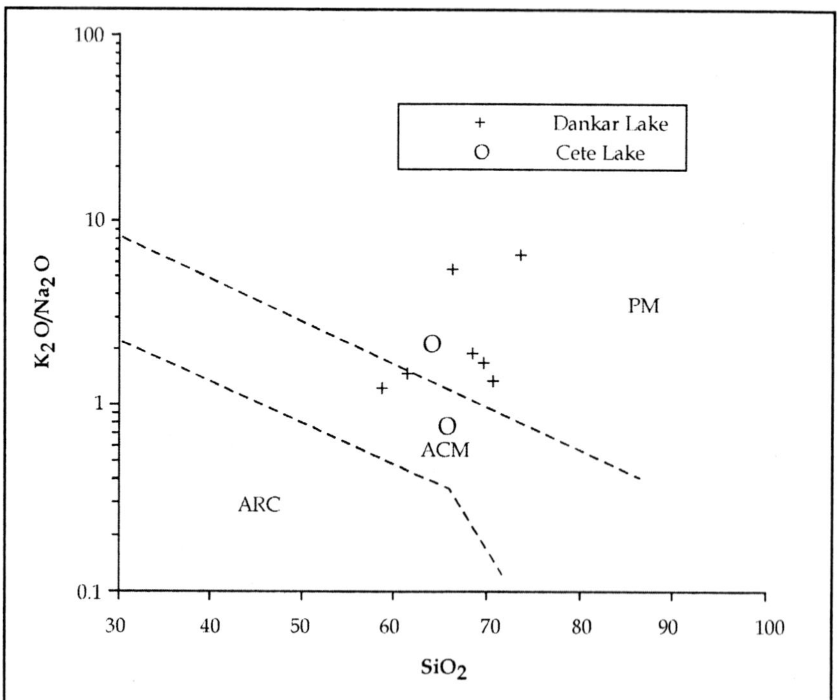

Fig. 7.21. Tectonic discrimination plot of spiti lake sediments which lie in passive margin (PM) and active continental margin (ACM) settings.

clay-shale rich samples plot in ACM field. In the study of modern sediment chemistry, Maynard and others (1982) used SiO_2 alone and presented sand and mud data for differing settings. On this diagram (K_2O/Na_2O vs SiO_2) same scenario emerges for the lake sediments.

❏❏❏

Chapter 8

BIOGEOCHEMISTRY OF LAKE SEDIMENTS

Organic matter constitutes a minor but important fraction of lake sediments. It is made up of a complex mixture of lipids, carbohydrates, proteins, and other biochemicals contained in the tissues of living benthic micro-organisms and contributed from the detritus of organisms formerly living in the lake and its water shed.

Significance of Sediment Organic Matter

The organic matter content of lake sediments participate in a variety of biochemical and geochemical processes. Benthic animals and microbes depend on organic matter for their nutrition. Because many components of organic matter are relatively easily oxidized forms of reduced carbon, the dissolved oxygen concentration of sediment pore water is controlled largely by the availability of oxidizable organic matter below the sediment/water interface. Further more, redox reactions, in which oxidized inorganic species become reduced by interaction with organic matter, are common in sediments organic matter, therefore, is a dynamic biogeochemical component of sediments.

Sources of organic Matter in Lake Sediments

The particulate detritus of plants that have lived in the waters and on the land surrounding a lake comprises the primary source of organic matter to the bottom deposits. Plants can be divided into two groups on the basis of their biochemical compositions: (1) non vascular plants that lack woody and cellulosic tissues, such as simple algae, and (2) vascular plants that have these tissues, such as grasses, shrubs, and trees. These latter types of plants exist on land and in shallow parts of lakes as bottom rooted emergent vegetation. Organic matter in lake sediments covers the spectrum of being predominantly algal in some lakes to being land-derived in others. The latter is probably more common in the lakes.

Bacteria and other microbes in the water and sediment of lakes and the soil of the lake watershed rework and degrade aquatic and land-derived organic matter. Although some bacteria are capable of photosynthesis or chemosynthesis, lake environments generally provide enough organic detritus. An important exception exists in strongly stratified lakes where bacteria can become major producers of primary organic matter.

In addition to these, local sources of organic matter and winds commonly transport material such as pollen, from sources outside the local watershed. The eolian component is typically a small fraction of the total organic mixture, yet it can contain distinctive components potentially useful to paleolimnetic and palaeoclimatic studies.

Depositional Processes Affecting Organic Matter

A variety of processes act to alter the character of organic matter in the relatively short time between its origin and its burial in the lake bottom. Foremost among these processes is its degradation during sinking. Studies show that nearly 85% of the carbon is oxidized before leaving the epilimnion, that part of the water column above the thermocline (Eadie et al., 1984). In more shallow depths, organic matter would have shorter sinking times and consequently, shorter exposures to oxidation in the water column. Shallow water lakes consequently, usually have sediments richer in organic matter than found in deep lakes like the Laurentian Great Lakes.

Bioturbation, or biological mixing of surface sediments, also prolongs the exposure of organic mater to oxidation and adds to this the degradation due to the nutritional needs of benthic fauna. The depth of bioturbation is generally low of the order of 10 cm or so. Many temperate-zone lakes experience periods of anoxia during summer thermal stratification, and the sediment of these lakes typically have high concentration of organic matter than do permanently oxygenated lakes. Varved sediments, in particular, indicate little or no resuspension and the absence of bioturbation. In general, components from aquatic sources are more sensitive to the bacterial degradative processes than are materials from land sources. Organic matter in lacustrine sediments consequently often have a large and relatively unreactive fraction of land-derived organic residues.

Due to multiple processes in the lake, the organic matter in sediments is often markedly different from that which is produced by the biota in and around the lake. Further more, the amount of organic matter buried in subsurface sediments is a small fraction of that originally introduced into the surface waters.

Indicators of Sources and Alterations of Total Organic Matter in lake Sediments.

Organic Carbon Stable Isotopes

Most photosynthetic plants incorporate carbon into organic matter using the C3 Calvin-Benson pathway which preferentially incorporates ^{12}C into organic matter, producing a shift of about –20 ‰ from the carbon isotope ratio of the inorganic carbon pool. Some plants use the C4 Hatch-Slack pathway, and others, mostly succulents, utilize the CAM pathway. The isotope shift for the C4 pathway is –8 to –12 ‰; that of the CAM pathway can vary from –10 to –20 ‰. Organic matter produced from atmospheric CO_2 by land plants using the C3 pathway consequently has an average $\delta^{13}C$ value of Ca-28 ‰ and by those using the C4 pathway Ca- 14 ‰ (O'Leary, 1988). Lake-derived organic matter is normally isotopically indistinguishable from organic matter from the surrounding watershed, except under special conditions.

The existence of the several isotopic signatures of photosynthetically fixed organic carbon would seem to offer opportunities to trace sources of organic matter in lacustrine food webs and to lake sediments. In the sedimentary organic matter of most lakes, however, the differences originally present in the total carbon isotopic contents of specific flora are lost because of food web integration to different carbon sources.

Microbial reworking of organic matter during early diagenesis can potentially modify its bulk carbon isotopic content. Organic matter is a mixture of different types of compounds which deviate differently in isotopic content from the inorganic carbon pool, Amino acids of marine plankton, for example, are depleted in 13 C by –17 ‰ on average, whereas the lipid fraction has a $\delta^{13}C$ value of –28 ‰ (Degens, 1969). Selective loss of more reactive fractions of the total organic matter may create a diagenetic shift in the isotope ratio which will mask the original source signature. The $\delta^{13}C$ ratio of the total organic carbon changes from –17 ‰ in surface sediments to –21 ‰ in sediments below 5m, a depth corresponding to Ca 4000 years ago, as organic carbon concentrations increase from 20 to 28%. This shift apparently is caused at least partly by selective loss of the isotopically heavy carbohydrate fraction of total organic matter (Spiker and Hatcher, 1984), but changes in the proportion of the land-derived and marine-derived organic matter may also contribute to it.

Nitrogen Stable Isotopes

Important paleolimnological information about organic matter source and diagenesis can be obtain from $\delta^{15}N$ measurements. Inorganic nitrogen reservoirs available to land plants and aquatic plants differ in their isotopic contents. Dissolved nitrate has an isotope ratio of +7 to +10 ‰, whereas atmospheric molecular nitrogen has a $\delta^{15}N$ of about 0 % (Peters et al., 1978). This difference is preserved in the isotopic contents of plankton ($\delta^{15}N$ of +8‰) and C3 land plants ($\delta^{15}N$ of +1 ‰) and has been used to trace food chain relationships in the coastal marine marsh by Peterson *et al.,* (1985).

The few investigations of the nitrogen isotope content of lake sediments show that the source information derived from the difference between atmospheric and dissolved forms of nitrogen appears to be preserved. In sediments from lake Superior, Ontario, Pang and Nriagu (1976, 1977) found that areas receiving large proportion of land-sourced organic matter have lighter $\delta^{15}N$ values than areas where aquatic organic matter is dominant (+ 3.7 ‰ vs + 4.9 ‰).

Organic Matter C/N Ratio

The presence or absence of cellulose in the plant sources of organic matter to lakes influence the C/N ratios of sediments. Nonvascular aquatic plants have low C/N rations, typically between 4 and 10, whereas vascular land plants which contain cellulose, have C/N ratios of 20 and greater. Lakes for which the contribution of organic matter from vascular land plants is small relative to water-column production, exemplified by Walker Lake and Lake Michigan, show lower C/N ratio in their sediments than do lakes receiving important amounts of vascular plant debris, such as Mangrove Lake and Lake Bosumtwi. Ratios of 13-14 for surface sediments of the latter two lakes suggest a mixture of nonvascular and vascular contributions, which is expected for most lakes. Studies carried out on various lakes have shown that selective diagenesis has no discernable effect on organic matter C/N rations in sediments though they are as old as 27,500 yr as in Lake Bosumtwi (Talbot and Johannessen, 1992). Therefore, a careful interpretation of C/N ratios can yield information about past sources of organic matter to lake sediments.

Diagenesis of Carbohydrates and Proteins in Lake Sediments

Carbohydrates and proteins are two important types of biologically synthesized organic matter. Carbohydrates encompass simple sugars which are water-soluble and reactive compounds and complex sugars having high molecular weights, which are consequently relatively insoluble and less reactive. The simple sugars are both produced and destroyed rapidly within sediments microcosms, with the result that the source and alteration pathways of these carbohydrate forms are difficult to decipher. Proteins, made up of amino acid units, can be similarly reactive. Interconversion between different amino acids is common again obscuring information about source and diagenesis. Both carbohydrate and protein compounds decrease relative to total organic carbon in sediments of Mangrove Lake and in general appear to decompose rapidly in lacustrine sediments (Kemp and Johnston, 1979, Hatcher *et al.*, 1982).

Paleolimnological Organic Geochemistry

The amounts and types of organic matter in lake sediments provide an integrated history of past life and conditions in and around each lake. Multidimensional organic geochemical studies have been done on sediments from a number of lakes. Indications of changes in the trophic status of these lakes and of changes in the watersheds around them have emerged from the organic matter records documented by these studies. Some of the lakes of Himalaya viz. Renuka, Rewalsar and Mansar have been investigated for their biogechemistry and the data and information are presented here.

Renuka Lake

Renuka is an oval shaped lake at an elevation of 620 m above msl in Lesser Himalaya having water spread of 670 ha. The lake follows a riparian course between two steep hill slopes with forests which are reported to have changed through time. According to Sharma (1992) at about 3000 yrs B.P, the overall dominance of oak and chirpine together with other broad-leaved texa have indicated warm and moist climatic conditions, whereas at about 2000-1000 yrs BP, decrease in oak and aquatic and increase in conifers and simultaneous decline in texa growing in wet conditions perhaps attributes to change in cool-dry conditions. Significant abrupt fall in oak, in tree and shrub pollen ratios demonstrates clearing of forests to bring more land under cultivation in the region around 500-600 yrs B.P.

The annual rainfall in this region is reported from 150-200 cm. The area is infested by major thrusts known as Krol-Giri thrust which has rendered geology highly complicated with development of folds, faults and crushing of rocks (Fig.2.4). The lake came into existence during Holocene period. A core of 65 cm length was raised from the lake for the biogeochemical and inorganic geochemical investigation (Das, 2001).

The lacustrine environment is a very complicated hydrological system, which is reflected in the isotopic composition of its sediments, especially the organic fraction. The main sources of organic carbon in a lake are submerged plants, such as macrophytes, floating pond weeds or planktonic material and terrestrial vegetation. Of these, submerged plants tend to be isotopically heavier because their source of CO_2 is dissolved bicarbonate rather than atmospheric CO_2. The enrichment of δ13C in the process $CO_2 \rightarrow HCO_3$, the latter being heavier, results in the fixed carbon also being enriched. Similarly, the δ13C value of planktonic material reflects the plankton source of carbon: use of dissolved bicarbonate results in 13C enrichment; use of atmospheric CO_2 leads to an overlap of planktonic δ13C with the values for terrestrial plants. The latter fall in into the well known C3 and C4 categories, with characteristic average δ13C values of –27 and -14 ‰, respectively.

Organic carbon $\delta^{13}C$ values change from –26.69 ‰ in near surface sediments to ca –21.45 ‰ at 61 cm depth and the organic carbon % also decreases below 17 cm depth. This shift indicates algal organic matter in near surface sediments which is isotopically light. The $\delta^{15}N$ isotopes are heavier in surface sediments 5.77 to 3.12 ‰ and lighter below 17 cm depth indicating land-sourced organic matter in bottom sediments. This reflects that the nitrogen isotope content of lake sediments, that gives source information, derived from the difference between atmospheric and dissolved forms of nitrogen isotope appears to be preserved. The C/N ratio of core sediments show higher values in near surface sediments while this value is low in the deeper sediments (Fig.8.1). This indicates a change in sediment types i.e. grain

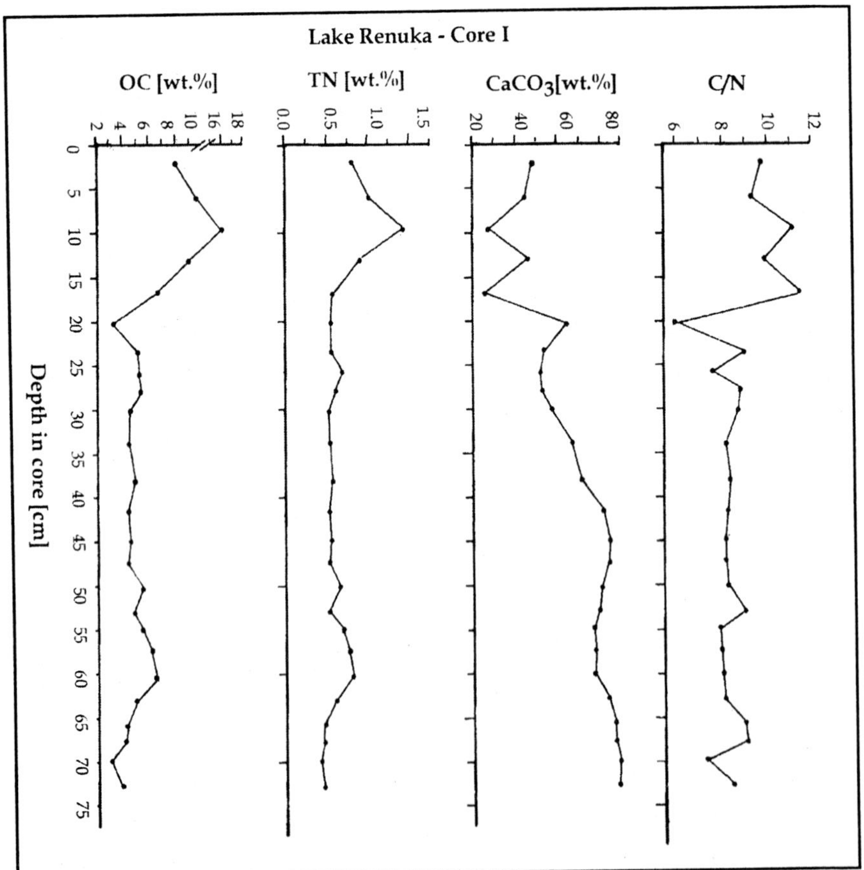

Fig. 8.1. Vertical variation in organic carbon (OC), total nitrogen (TN), $CaCO_3$ and C/N ratio in Renuka lake core sediments.

size variation from fine-grained in deep water to sandy in shallow- water and /or mixture of vascular and non-vascular organic matter.

Rewalsar Lake

A 130 cm core was raised from this lake. The $CaCO_3$ content in the core sediment is low with a tendency to decrease down wards with an exception at 14, 49, 72 and 89 cm depth. The lake is undersaturated in $CaCO_3$ with more of dissolved CO_2 in water (Das and Kaur, 2001). Lower organisms such as algae, phytoplankton and zooplankton are characterized by a high protein content, hence low C/N ratios, while terrestrial plants which are the main source of terrestrial organic matter, are characterized by low protein content hence high C/N ratios of bottom sediments and can be used as a powerful complement to their $\delta^{13}C$ values in evaluating the relative proportions of autochthonous and allochthonous carbon sources.

The organic carbon $\delta^{13}C$ is lighter at a very shallow depth~ -33.5 ‰ and then becomes relatively heavier ranging between -28 ‰ to -23 ‰, but in deeper sediments at about 74 cm depth, it shows increasing trend becoming heavier and reaches -20 ‰ indicating increase in non-vascular sources of organic carbon. In comparison to this, the C/N ratio in shallow sediments ranges from 9.75 to 11.75 and in deeper sediments it shows decreasing trend, and after 74 cm depth it reaches a minimum of 8.5 (Fig. 8.2).

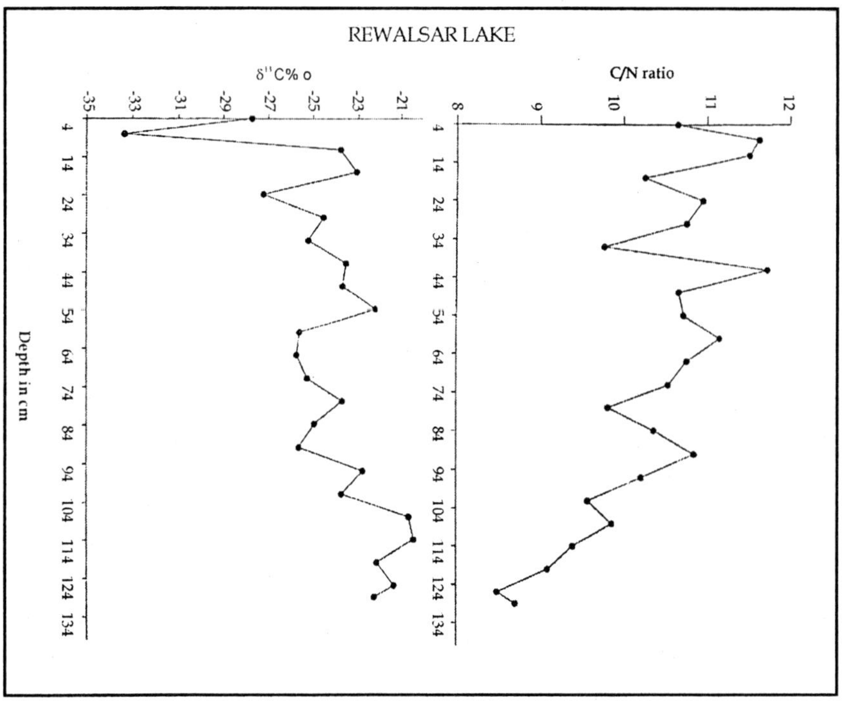

Fig. 8.2. Vertical distribution of δ13C and C/N ratio in Rewalsar Lake sediments.

A sort of anti-correlation is exhibited indicating different sources of organic carbon in the sediment. The lower C/N ratio indicates non-vascular source of organic matter enriched in protein content. The increasing δ13C values suggest aquatic plants and planktonic material deriving carbon from dissolved bi-carbonate in deeper sediments, enriching 13C carbon values. Therefore, the δ13C values indicate greater productivity in deeper lake during wet and hot climatic regime where as in shallower sediments C/N ratio and δ13C suggests terrestrial sources of organic carbon in sediments. The data, therefore, indicate wet and warmer climatic regime during early Holocene and cold and dry conditions in the late Holocene time. This compares with the findings from other lakes such as Spiti region lakes and lake sediments in Tibet, though several warm and cold events have also been recorded during Holocene in the Himalayan region.

Mansar Lake

The Mansar Lake in Lesser Himalaya, Jammu, is the deepest (38.25 m) fresh water lake in the region. It is a tectonic lake and came into existence during Holocene Period. The catchment of the lake is composed of fine-grained sandstones alternating with siltstones, mudstones and clays of the Lower Siwalik Formation. The average annual rainfall is 1500 mm and there is no perennial stream flowing into the lake from the catchment. The lake is thus fed essentially by rain water and springs underground.

A long core of 30 m length was raised from this lake and the same has been used for inorganic geochemistry as well as bio-geochemical investigations. The organic carbon show little variation at shallow levels up to 16 m depth, and from 16 m to 21.9 m flat profile, is observed, and then marked increase and decrease up to 30 m depth (Fig.8.3). The flat profile is the observed, coarse sand in the samples. The total nitrogen shows almost the same pattern i.e. little variation at shallow depth but steep increase at depth beyond 21 m to 30m. The C/N ratio also shows increase from 22.29 m to 30 m depth (Fig.8.3).

The $\delta^{13}C$ graph of Mansar Lake shows that at a shallow depth of about 6 m δ13C are relatively heavier and it gradually depletes and becomes lighter till 16 m depth. The heavier 13C at very shallow depth may be due to algae or floating pond weed etc (Fig.8.3). The C/N ratio at shallow depth range from 6-13 with a variation between 8 to 11 and this value is not typically similar to non-vascular aquatic plants because this value is rarely reported to be more than 10. The δ13C values of –25 to -23 ‰ support the contention that the relative depletion in δ13C could be due to a greater contribution from terrestrial vegetation. This may indicate autochthonous and allocthonous carbon sources. In deeper sediments δ13C -20 ‰, -21 ‰, and -22 ‰ corresponds to the depths and C/N ratio more than 16 reaching 20 (Fig. 8.3). However, δ13C values deviate from the extreme average value of -17 ‰ expected of aquatic plants

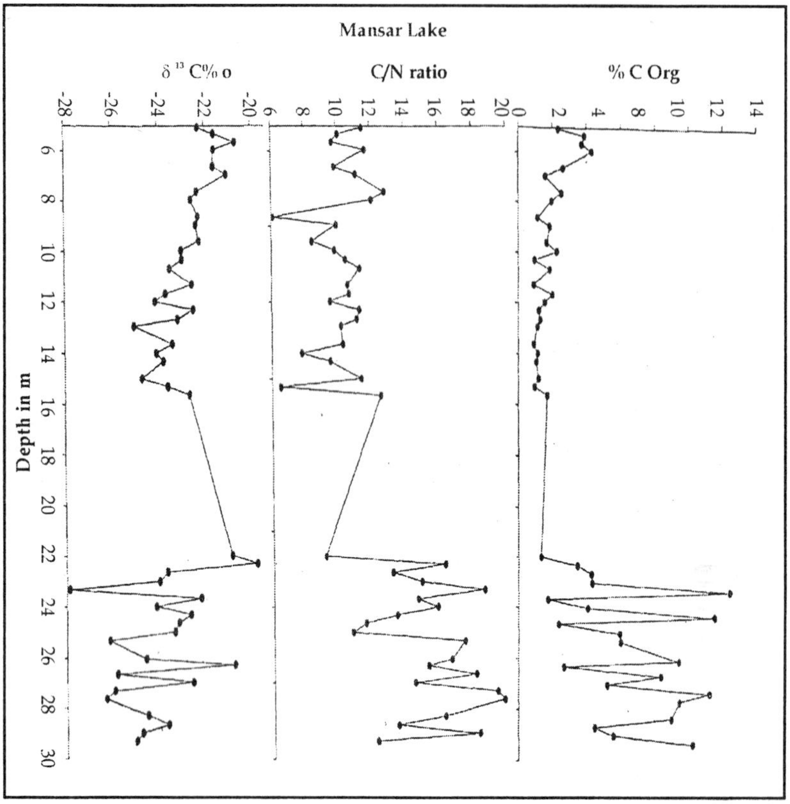

Fig. 8.3. Vertical distribution δ13C, C/N ratio and organic carbon in sediments of Mansar Lake.

using dissolved bicarbonate. This lowering of δ13C values may be due to a decrease in supply of molecular CO2 caused by more flushing and a relative increase in floating planktonic material. The total organic carbon content is lower in shallow sediments and in deeper sediments its % is higher indicating greater productivity during warm-wet conditions resulting in an increased carbon content of total sediment and more dissolved bicarbonate contributed to the lakes carbon pool to keep pace with the requirements of a highly productive lake. The study carried out on lake Biwa sediments, Japan has also shown similar good correlation between δ13C values of the organic fraction and the absolute carbon content when the lake was highly productive during warm period. During a cold phase the opposite effect operated, making the organic carbon relatively depleted in 13C and also causing a decrease in the absolute carbon content. This shows warmer and wetter climatic regime during early Holocene followed by enhanced precipitation when predominantly sands were eroded and deposited and then the climate ameliorated to cold and dry conditions during late Holocene.

REFERENCES

Abbas. N., Subramanian V. 1984. Erosion and sediment transport in the Ganges river basin, India. Jour. Hydrol., 69, 173-182.

Ahmed, S. AL-Mikhlafi, Das, B.K., Kaur, P., 2003. Water chemistry of Mansar Lake (India): an indication of source area weathering and seasonal variations. Env. Geol. 6, 1-9.

Allen, H.F., and Kramer J.R., 1972. Nutrients in Natural waters. John Wiley-Interscience, New York, NY.

Alexander, M., 1961. Introduction to soil microbiology, John Wiley, New York, 472 pp.

Appleby, P.G., and Oldfield,F., 1992. Application of lead-210 to sedimentation studies. Ivanovich and R.S.Harmon (Eds.), Uranium-Series Disequilibria; Applications to Earth, Marine, and Environmental Sciences, 2nd ed., pp. 731-778. Clarendon, Oxford, UK.

Arogyaswamy, R.N. P., 1962. The Lonar lake. Ind. Min., 169-11.

Bauluz, B., Mayayo, M.J., Fernandez-Nieto, C. and Lopez, J.M.G., 2000. Geochemistry of Precambian and Paleozoic siliciclastic rocks from the Iberian Range (NE spain): implications for source area weathering, sorting, provenance, and tectonic setting. Chem. Geol., 168, 135-150.

Basu, A., Blanchard, D.P., Brannon, J.C., 1982. Rare-earth elements in the sedimentary cycle: a pilot study of the first lag. Sedimentology pp 737-742.

Bea, F., 1996. Residence of REE, Y and Th in granites and crustal protoliths, implications for the chemistry of crustal melting. Jour. Petro., 37, 521-552.

Beals, C.S., Innes, M.J.S., Rottenburg, J.A., 1960. The search for fossil meteorite craters. curr. sci. 29, 205-217 and 249-260.

Benninger, L.K., Aller, R.C., Dion, E.P., Turekian K.K., 1977. Pb^{210} distribution in sediments of Long Island Sound. Trans. Am. Geophys. Union, 58, 422.

Bhatia, M.R., 1983. Plate Tectonics and geochemical composition of sandstones. Jour. Geol., 91, 611-627.

Berner, E.K., and Berner, R.A., 1987. The Global Water cycle. Prentice-Hall, Englewood Cliffs, N.J., 397 pp.

Berner, E.K., 1971. Principles of chemical sedimentology. McGraw-Hill, New York, 240pp.

Berner, R.A., 1968. Calcium carbonte concretion formed by the decomposition of organic matter. Sci. 159, 195-197.

Bhattacharya, B.N., Gupta, S.K., Rishi, B.P., Mitra, A.K., Kashyap, V.K., 1982. Seasonal variation in the composition of Sambhar Lake brine and possible origin of its salinity. Misc. Publ.-Geol. Surv., India, 49, 80-89.

Bortleson, G.C., and Lee, G.F., 1972. Recent sedimentary history of lake Mendota, Wis. Environ. Sci. Tech., 9, 799-808.

Cardenas, A., Girty, G.H., Hanson, A.D., Lahren, M.M., 1996. Assessing differences in composition between low metamorphic grade mudstones and high-grade schists using logratio techniques J. Geol. 104, 279-293.

Chaudhuri, S., and Cullers, R.L., 1979. The distribution of REE in deeply buried Gulf Coast sediments. Chem. Geol. 24, 327-338.

Chawdhury, A.N., and Handa, B.K., 1978. Some aspects of the Geochemistry of Lonar Lake water, Ind. Jour. Earth Sciens., 5, 1, 111-118.

Correns , C.W., 1949. Einfuhrung in die Mineralogie. Springer-Verlag, Berlin. 414 pp.

Condie, K.C., 1991. Another look at rare earth elements in shales. Geochim. et Cosmochim. Acta. 55, 2517-2531.

Condie, K.C., 1993. Chemical composition and evolution of the upper continental crust: contrasting results from surface samples and shales. Chem. Geol. 104, 1-37.

Covey, M., 1986. The evolution of foreland basins to steady state evidence from the Western Taiwan foreland basin. In : Allen, P.A., and Homewood, P. (eds.) Foreland Basin special publication of International Association of Sedimentology. 8, 77-90.

Cox, R., Lowe, D.R., Cullers, R.L., 1995. The influence of sediment recycling and basement composition on evolution of mud rock chemistry in the South-Western United States. Geochim. et Cosmochim. Acta, 59, 2919-2940.

Craig, R.J., 1972. Pollen influx to laminated sediments: a pollen diagram from North-Eastern Minnesota. Ecology. 53, 46-57.

Cullers, R.L., 1994. The chemical signature of source rocks in size fractions of Holocene Stream sediment derived from metamorphic rocks in the wet mountains region. Colorado, USA, Chem. Geol. 113, 327-343.

Cullers, R.L., Barrett, T., Carlson, R., Robinson, B., 1987. REE and mineralogic changes in Holocene soils and stream sediments. Chem. Geol, 63, 275-297.

Cullers, R.L. Graf, J., 1983. Rare earth elements in igneous rocks of the continental crust: intermediate and silicic rocks, ore petrogenesis. In: Henderson, P. (Ed.), Rare Earth Geochemistry. Elsevier. Amsterdam, 275-312.

Cullers, R.L., Stone, J., 1991. Chemical and Mineralogical composition of the Pennsylvanian mountain, Colorado, USA an uplifted continental blocks to sedimentary rocks from other techtonics environments. Lithos 27, 115-131.

Das, B.K., 1999. Environmental Pollution of Udaisagar Lake and impact of phosphate mine, Udaipur, Rajasthan, India. Env. Geol. 38, 3, 244-248.

Das, B.K., 2001. Ecohydrology and Biogeochemisty of the Himalayan Lakes: A case study. Ecohydrology, (eds.) Subramanian and Ramanathan, Proc. UNESCO sponsored workshop, Capital Publishing Co., New Delhi. 27-41.

Das, B.K., 2005. Environmental Pollution impact on water and sediments of Kumaun Lakes, Lesser Himalaya, India: a comparative study. Env. Geol., 49, 230-239.

Das, B.K., AL-Mikhlafi, Kaur, P., 2006. Geochemistry of Mansar Lake sediments, Jammu, India: Implication for source-area weathering, provenance, and tectonic setting. Jour. Asian Earth Sciences, 26, 649-668.

Das, B.K., Birgit-Gaye Haake, 2003. Geochemistry of Rewalsar Lake sediment, Lesser Himalaya, India: implications for source-area weathering, provenance and tectonic setting. Geosciences Journal, 7,4,299-312.

Das, B.K., and Dhiman, S.C., 2003. Water and Sediment Chemistry of Higher Himalayan lakes in the Spiti Valley: Control on weathering, provenance and tectonic setting of the basin. Env. Geol., 44, 717-730.

Das, B.K., and Kaur, P., 2001. Major ion chemistry of Renuka Lake and weathering processes, Sirmaur Distt., Himachal Pradesh, India. Env. Geol. 40, 7, 908-917.

Das, B.K., Kakar, R.K. 2000. Major ion chemistry and weathering processes of Mansar Lake, sub- Himalaya, Jammu, India. Jour., Punj. Acad. Scien, 2 (i) 241-247.

Das B.K., Singh, M., Borkar, M.D. 1994. Sediments accumulation rate in the lakes of Kumaun Himalaya, India, using Pb^{210} and Ra^{226}. Env. Geol. 23, 114-118.

Das, B.K., and Singh M. 1994. Rate of sediment accumulation in lakes of Udaipur, Rajasthan, India, using the Pb^{210} method. Env. Geol., 24, 28-33.

Das, B.K., and Singh M. 1996. Water chemistry and control of weathering of Pichola Lake, Udaipur district, Rajasthan, India. Env. Geol., 27, 184-190.

Das, S.M., Pandey, J., 1978. Lake Pollution as evidenced by physical, biological and chemical parameters in the Nainital Lake, U.P., Proc. Nat. Sem. Resources Development and Environment in the Himalaya Region, Deptt. Of Science and Technology, New Delhi, 502-512.

Degens, E.T., 1969. Biogeochemistry of stable carbon isotopes. In Organic Geochemistry-Methods and Results (Edited by Eglinton G. and Murphy M.T.J., 304-329. Springer, New York.

deGaridel-Thoron, T., Beaufort, L.,Lonsley, B.K. and Dannenmann's, 2001. High-resolution temperature and evaporation changes during the late Holocene in the north-eastern Arabian Sea. Paleoceanography. 16(4), 358-367.

Drever, J.I., 1982. The Chemistry of Natural Works. Englewood Cliffs, N.J. Prentice Hall, 388 pp.

Drever, J.I., 1988, The Geochemistry of Natural Waters. Prentice Hall, p. 437.

Drever, J.I., 1997. The Geochemistry of Natural Works, 3rd ed. Prentice-Hall, 436 pp.

Durham, R.W., and Joshi, S. R., 1980. The Pb^{210} and Cs^{137} profiles in sediment cores from Lakes Matagami and Quevillon , North-West Quebec, Canada: Canadian Journal of Earth Science, 147, 1746-1750.

Eadie, B.J., Chambers, R.L., Gardner, W.S., Bell, G.E., 1984. Sediment trap studies in Lake Michigan: Resuspension and chemical fluxes in the Southern basin. Jour. Great Lake Res., 10,307-321.

Eugster, H.P. Hardie, L.A., 1978. Saline Lakes. In : Lerman A. (Ed.), Lakes: Chemistry, Geology, Physics, Springer-Verlag, pp. 237-293.

ENEX 1978: Study of the Pollution of Dal Lake, Srinagar, Kashmir-Enex Consortium, Newzeland.

Fleet, A.J., 1984.Aqueous and sedimentary geochemistry of the rare earth elements. In: Rare Earth Elements Geochemistry, P.Henderson (Ed.), Elsevier, Amsterdam.Cap 10:343-373.

Förstner, U., Wittmann G.T.W., 1979. Metal Pollution in the Aquatic Environment. Springer-Verlag, Berlin, 486 p.

Fredricksson, K., Dube, A., Milton, D. and Balasundram, M.S., 1973. Lonar Lake, India-An impact crator in Deccan trap. Science, 180, 862-864.

Fredricksson, K. Brenner, P., Dube, A., Milton D, Mooring, C. and Nelen J.A., 1979. Petrology, mineralogy, and distribution of Lonar (India) and lunar impact breccias and glasses, Smithsonian Contributions to the Earth Sciences, 22, 1-12.

Fudali, R.R., Molton, D.J., Fredriksson K., Dube, A. 1980. Morphology of Lonar crater, India: comparison and implications. The Moon and Planets 23, 493-515.

Gahler, A.R. 1969. Sediment-water nutrient interchange. In: Eutrophication, Biostimulation Assessment Workshop. U. of California, Berkeley, C.A.

Gale A, France-Lanord, C., 1999. Weathering processes in the Ganges-Brahmaputra basin and the riverine alkalinity budget, Chem. Geol. 159, 31-60.

Garrels, R.M., 1967. Genesis of some groundwater from igneous rocks. Researches in Geochemistry, Vol. 2, P.H. Abelson ed. Wiley, New York 420 pp.

Garrels R.M., Christ, C.L., 1965. Solutions, minerals and equilibria. Freeman Cooper, San Francisco.

Garrels R.M., Mackenzie, F.T., 1967. Origin of the chemical compositions of some springs and lakes. Equilibrium concepts in Natural Water Systems. Am. Chem. Soc. Adv. Chem. Ser.67, 222-242.

Garrels, R.M., Mackenzie, F.T., 1971. Evolution of sedimentary rocks. W.W. Norton, New York, NY, 397p.

Garrels R.M., Perry, E.A., 1974. Cycling of carbon, sulphur, and oxygen through geologic time. In the sea (ed. E.D. Goldberg) 5, 303-336., Wiley Interscience, New York.

Gibbs, R.J., 1970. Mechanism controlling World Water Chemistry. Science 170, 1088-1090.

Gibbs, R.J., 1977. Clay mineral segregation in the marine environment. Jour. Sed. Petrol. 47 (1), 237-243.

Graham, W.F., Duce, R.A. 1979. Atmospheric pathway of the phosphorus cycle,Geochim. et. Cosmochim. Acta. 43, 1195-1208.

Graver, J.I., Scott, T.J., 1995. Trace elements in shale as indicators of crustal provenance and terrian accretion in south Canadian Cordillera. Geol. Soc. Am. Bull. 107,440-453.

Gromet, L.P., Silver, L.T., 1983. Rare-Earth Element distribution among minerals in a granodiorite and their petro-genetic implications. Geochimica et Cosmochimica Acta, 47, 925-940.

Grim, R.E., 1968. Clay Mineralogy; 2nd ed., McGraw-Hill, New York, 596 p.

Godbole, N.N., 1952. The salinity of Sambhar Lake, Bull. Natl. Inst. Sci. India, 1, 89-93.

Goldich, S.S., 1938. A Study of rock weathering J.Geol. 46, 17-58.

Goldhaber, M.B., Aller, R.C., Cochran, J.K., Rosenfeld, J.K., Marten's, C.S., Berner, R.A., 1977. Sulfate reduction, diffusion and bioturbation in Long insland, sound sediments. Report of the FOAM group. 277, 193-237.

Handa, B.K., Kumar, A., Geol, D.K., 1982. Eutrophication of Nainital Lake. IAWPC Tech. Manual IX, 110-120.

Hardie, L.A., Eugster, H.P., 1970. The evolution of closed basin brines. Min. Soc. Am. Spl. Pap. 3,273-290.

Haskin, L.A., Wilderman, R.T., Frey, F.A., Collins, K.A., Keidy, C.R., and Haskin, M.A., 1966. Rare earths in the sediments. Jour., Geophys. Res., 71, 6091-6105.

Hatcher, P.G., Simoneit, B.R.T., Mackenzie, F.T., Neumann, A.C., Thorstenson, D.C., and Gerchakov, S.M., 1982. Organic geochemistry and pore water chemistry of sediments from Mangrove Lake, Bermuda. Org. Geochemistry. 4,93-112.

Heron, A.M., 1953. The geology of Central Rajputana. Mem, Geol., Surv. India, 79, 389 pp.

Herron,M.M.,1988. Geochemical classification of terrigenous sands and shales from core or log data. Jour. Sed. Petrol. 58, 820-829.

Holland H.D., 1978. The chemistry of oceans and atmosphere. Wiley, New York.

Holland, T.H., Christie, W.A.K., 1909. The origin of the salt deposits of Rajputana. Rec. Geol. Surv. India 38(2), 154-186.

Hutchinson, G.E., 1957. A treatise on limnology (1) John Wiley, New York, 1015 pp.

Hutchinson, G.E., 1937. Limnological Studies in India. Tibet. Int. Rev. Gesamten Hydrobiol. 35, 134-177.

Jauhari, P. Hashimi, N.H., 1994. On the surficial sediments of the fresh water Naini Lake, Kumaun Himalaya, India. Proc. Indian National Sci. Acad. , 60, 5, 675-682.

Jenne, E.A. 1976. Trace element sorption by sediments and soils sites and processes. Sym. On Molybdenum in the Environment, Vol.2, (eds. W. Chappel and K.Peterson). Marcel Dekkar, New York, 553 pp.

Jones, F., Bowser, C.J. 1978. The mineralogy and related chemistry of lake sediments. In: A Lerman (Ed.), Lakes, chemistry, Geology, and Physics, Springer, New York 179-227.

Kanth, T.A., Rather, G.M, 2001. Sedimentation and shrinkage of Dal Lake-Kashmir, In: Subramanian, V., Ramanathan, A.L. (eds.) Ecohydrology, Proceedings of the International workshop on Ecohydrology, Delhi, India, 9-16.

Kemp, A.L.W., Anderson, T.W., Thomas, R.L., and Modrochova, A. 1974. Sedimentation rates and recent sediment history of Lakes Ontario, Erie and Huron, J. Sed. Petrol. 44,207-218.

Kemp, A.L.W., Johnston L.M., 1979. Diagenesis of organic matter in the sediments of Lakes Ontario, Erie and Huron. J. Great Lakes, Res. 5, 1-10.

Khan, M.A., Zutshi, D.P., 1980. Contribution to high altitude limnology of Himalayan system. I.Limnology and primary productivity of plankton community in L.Nilnag, Kashmir Hydrobiologia, 75:103-112.

Khan, M.A., Zutshi, D.P., 1992. Studies on evaluation of energy flux at primary producer level in three lakes of Kashmir, India. The Japanese Journal of Limnology, 53, 3,217-221.

Koide, M. Bruland, K.W., Goldberg, E.D., 1973. Th-228/Th-232 and Pb^{210} geochronologies in marine and lake sediments, Geochim. et. Cosmochim. Acta. 37, 1171-1187.

Krauskoff, K.B. 1955. Sedimentary deposits of rare metals. Econ. Geol. 50th Ann. Vol., 413-463.

Krishnaswami, S., Lal. D, Martin, J.M., Meybeck, M. 1971. Geochronology of lake sediments. Earth Planet. Sci. Lett. 11, 407-414.

Krishnaswami, S., Lal, D. 1978. Radionuclides limnochronolgy. In: A Lerman (Ed.), Lakes-Chemistry, Geology and Physics. Spinger, New York, Ny, 153-177.

Kundrass, H.R., Hofmann, A., Doose, H., Emeis, K.C. and Erlenkeuser, H., 2001 Modulation and amplification of climatic changes in Northern Hemisphere by the Indian summer monsoon during the past 80 Ky. Geology 29 (1) 63-66.

Kundangar, M.R.D., Sarwar, S.G. 1997. Dal Lake Kashmir. Publ. by hydrobiology research laboratory, Kashmir, 37p.

Lafond, E.C. and Dietz, R.S. 1964. Lonar Crater, India, a meteorite crater ? Meteoritics 2, 111-116.

Lederer, C.M., Hollander, J.M., and Perlman, I, 1967. Table of isotopes. 6th ed., John Wiley and Sons, New York.

Lerman, A. and Brunskill, G.J. 1971. Migration of major constituents from lake sediments into lake waters and its bearing on lake water composition. Limnol. Oceanogr., 16, 880-890.

Löffler, H., 1969. High altitude lakes in Mt.Everest region-Verh. Internat. Verein. Limnol. 17, 373-385.

Loughnan, F.C., 1969. Chemical weathering of silicate minerals. New York: American Elsevier. 154 pp.

Maher, L.J., Jr. 1977 Palynological Studies in the western- arm of lake Superior. Quat. Res., 7, 14-44.

Manheim, F.T., 1976. Interstitial Waters of marine sediments. In: JP. Riley and Chester, R (eds.) chemical oceanography. 2nd ed. Vol. 6, 115-186.

Marsh, W.M., Dozier, J., 1981. Landscape: An Introduction to Physical Geography. Addison-Wesley, Reading, 637 p.

Martin, J.M., Hogdahl, O. Phillpott, J.C., 1976. Rare earth element supply to the ocean. J. Geophys. Res. 81, 3119-3121.

Mason, B., 1960. Principles of Geochemistry. John Wiley & Sons, INC 310 pp.

Matthess,W.,Frimmel,F.,Hirsch,P.,Schulz,H.D.,Usdowski,H.E.(Eds.).1992. Progress in Hydrogeochemistry. Springer-Verlag.

Maynard, J.B.,Valloni, R., Yu, H.S., 1982. Composition of modern deep-sea sands from arc-related basins. Geol. Soc. Lond spec. Publ. 10, 551-561.

McLennan, S.M., 1984 Petrological characteristics of Archean greywackes, Jour. Sed. Petrol., 54, 889-898.

McLennan, S.M., 1989. Rare earth elements in sedimentary rocks: Influence of provenance and sedimentary processes. Review Mineralogy, 21, 170-199.

McLennan, S.M., Hemming, S., McDaniel, D.K., Hanson, G.N., 1993. Geochemical approaches to sedimentation, provenance and tectonics. In: Johnsson, M.J. Basu A (eds.) Processes controlling the composition of clastic sediments. Geol. Soc.Am Special Paper, 284, 21-40.

McLennan, S.M., Taylor, S.R., Kroner, A 1983. Geochemical evolution of Archaen Shales from South Africa. I, The Swaziland and Pongola Supergroups. Precambrian Res., 22, 93-124.

McLennan, S.M, Taylor, S.R., McCulloch, M.T., Maynard, J.B., 1990. Geochemical and Nd-Sr isotopic composition and plate tectonics associations. Geochim et. Cosmochim Acta 54, 2015-2050.

McLennan, S.M., Taylor, J.R., 1991, Sedimentary rocks and crustal evolution: Tectonic Setting and Secular Trends. Jour. Geol. 99, 1-21.

Meybeck, M., 1979. Concentration des eaux fluviales en elements majeurs et apports en solution aux oceans. Rev. Geol., Dyn., Geogr, Phys. 21(3), 215-216.

Meybeck, M., 1980. Pathways of major elements from land to ocean through rivers. In : J.M. Martin, J.D. Burton and D. Eisma (Eds.) Proc. of the review and workshop on river inputs to ocean-systems, Rome, 18-30.

Meybeck, M., 1982. Carbon, nitrogen and phosphorus transport by world rivers, Am. Jour., Sci., 282, 401-450.

Meybeck, M., 1984. Les fleuves et le cycle geochimique des elements. These d'etal (no 84-35). E'cole Normal Superieure, Laboratoire de Geologie, Univ. Pierre et Marie curie, Paris 6, France.

Meybeck, M. 1987. Global chemical weathering of surficial rocks estimated from river dissolved loads. Am. Jour. Sci. 287, 401-428.

Middlebrooks, E.J., Falkenberg, D.H., Maloney, T.E., 1974. Modelling the Eutrophication Process. Ann. Arbor. Science. Pub., Ann. Arbor, MI.

Millot, G. 1970. Geology of Clays. Springer-Verlag, Paris, 305pp.

Moore, C.A., Silver, M.L. 1975, Nutrient transport by sediment- water interaction. Proc. Int. Clay. Conf. , 495-504.

Murali, A.V., Parthasarathy, R., Mahadevan, T.M., and Shankar Das, M., 1983. Trace element characteristics, REE patterns and partition coefficients of zircons from different geological environments-A case study on Indian zircons. Geochim. et Cosmochim. Acta. 47, 2047-2052.

Nance, W.B., Taylor, S.R., 1976. Rare earth patterns and crustal evolution I: Australian post-Archean, sedimentary rocks. Geochim et cosmochim Acta, 40, 1539-1551.

Nandy, N.C., Deo, D.B., 1961. Origin of Lonar lake and its alkalinity, TISCO, 8, 144-155.

Nayak, V.V., 1972. Glassy objects (Impactite glasses), a possible new evidence for meteoritic origin of the Lonar Crater. Earth and Planetary Sci. Lett. , 14, 1-6.

Negrel, P, Allegre, C.J. Duper, B., Lewin, E., 1993. Erosion sources determined by inversion of major and trace-element ratios in river water: the Congo Basin case. Earth Planet Sci. Lett. 120, 59-76.

Nesbitt, H.W., 1979. Mobility and fractionation of rare earth elements during weathering of a granodiorite. Nature, 279, 206-210.

Nesbitt, H.W., Markovics, G., Price, R.C., 1980. Chemical processes affecting alkalies and alkaline earths during continental weathering. Geochim. et. Cosmochim. Acta, 44, 1659-1666.

Nesbitt, H.W., Young, G.M., 1982. Early Proterozoic climates and plate motions inferred from major element chemistry of lutites. Nature, 299, 715-717.

NRSA., 1996. IRS-IC LISS-III, Imagery, Dated 27-5-1996.

Nygaard, G., 1949. Hydrobiological Studies on some Danish ponds and lakes. Part-2. The quotient hypothesis and some new or little known phytoplanktonic organisms. Det. Kongelliga Danske, Vindeskabernes Selskab. Biologiska Shifter. 7, 1-293.

O'Leary, M.H., 1988. Carbon Isotopes in Photosynthesis. Bioscience, 38, 328-336.

Pande K., Sarin, M.M., Trivedi, J.R., Krishnaswami, S., Sharma, K.K., 1994. The Indus river system (India-Pakistan): Major-ion chemistry, Uranium and strontium isotopes. Chem. Geol. 116, 245-259.

Pang, P.C., Nriagu, J.O., 1976 Distribution and isotope composition of nitrogen in Bay of Qinte (Lake Ontario) sediments. Chem. Geol., 18, 93-105.

Pang, P.C., Nriagu, J.O., 1977. Isotopic variations of the nitrogen in Lake Superior. Geochim. et Cosmochim. Acta, 41, 811-814.

Pant, M.C., Sharma, A.P., Sharma, P.C., 1980. Evidence for the increased eutrophication of Lake Nainital as a result of human interference. Env. Poll. Ser. B: 149-161.

Pant., M.C., Sharma, A.P., 1978. Observations on some possible causal factors governing the mortality of winter fish in the Nainital Lake. In: Proceedings of the National Seminar on Resource Development and Environment in the Himalayan Region, DST, New Delhi, 248-254.

Pennington, W., Cambray, R.S., Fisher, E.M., 1973. Observation on lake sediments using fallout 137Cs as a tracer. Nature, 242, 324-326.

Pettijohn, F.J., Potter, P., Siever, R. 1972. Sand and sandstone. Springer-Verlag, New York.

Peterson, B.J., Howarth, R.W., Garritt, R.H., 1985. Multiple stable isotopes used to trace the flow of organic matter in estuarine food webs. Science. 227, 1361-1363.

Peters, K.F., Sweeney, R.E., Kaplan, I.R., 1978. Correlation of carbon nitrogen stable isotope ratios in sedimentary organic matter. Limonol. Oceans org. 23, 598-604.

Potter, P.E., 1978. Petrology and chemistry of big river sands. Jour. Geol., 86, 423-449.

Prescott, W.G., 1968. The Algae. Houghton Mifflin Col. Boston. Mass., 1-314.

Rai, S.P., Kumar, V., Singh, O., Kumar, B. and Jain, S.K., 2006. Bathymetry, sedimentation rate and physico-chemical characteristics of Mansar Lake in the Himalayan Foothills, Jammu and Kashmir, India. Jour. Geol. Soc. Ind. 67, 211-220.

Raina, B.N. Dungrakoti, B.D., 1975. Geology of the area between Nainital and Champawat, Kumaun Himalaya, U.P, Him. Geol. 5, 1-28.

Rawat, J.S., 1987. Morphometry and Morphology of the Nainital Lake. Jour. Geol. Soc. Ind., 30, 493-498.

Ramesh, R., Jani, R.A., Bhushan, R., 1993. Stable isotopic evidence for the origin of water in salt lakes of Rajasthan and Gujarat, Jour. Arid Environ. 25, 117-123.

Rajasthan Pollution Prevention and Control Board 1986. Environmental status of lakes in Rajasthan. Environmental status document series on surface waters ESDS, SWI, 86-87.

Rastogi, S.P., 1979. Geochemical Characteristics and Evolution of the Tso-Kar lake, Geol. Sur. India, Misce. Publ. No. 41, Pt I. 259-278.

Raymahashay, B.C., 1986. Geochemistry of bicarbonate in river water, Jour. Geol., Soc. India, 27, 114-118.

Rollinson, H.R., 1993. Using Geochemical Data: Evaluation, Presentation, Interpretation, Longman, 352p.

Roser, B.P., Korsch, R.J., 1986. Determination of tectonic setting of sandstone-mudstone suites using SiO_2 content and K_2O / Na_2O ratio. J. Geol. 94, 635-650.

Roser, B.P., Korsch, R.J., 1988. Provenance Signatures and Sandstone-mudstone suites determined using discriminant function analysis of major-element data. Chem. Geol. 67, 119-139.

Roy., A.B. 1988. Stratigraphic and tectonic framework of the Aravalli Mountain Range. Geol., Soc., India, Mem 7, 3-31.

Roy, P.D., Sinha, R., Smykatz-Kloss, W., 2001. Mineralogy and geochemistry of the evaporitic crust from the hypersaline Sambhar Lake playa, Thar Desert, India. Chemie der Erde (Geochemistry) 61, 241-253.

Rosen, M.R. 1991. Sedimentologic and Geochemical constraints on the hydrologic evolution of Bristol Dry lake, California, U.S.A. Paleogeogr. Paleoclim. Paleoecol. 84, 229-257.

Rosen M.R., 1994. The importance of groundwater in playas, a review of playa classifications and the sedimentology and hydrology of playas. Geol. Soc. Am. Spl. Pap. 289, 1-18.

Sarin, M.M., Krishnaswami, S., Dilli, K., Somayajulu, B.L.K., Moore, W.S. 1989. Major ion Chemistry of the Ganga-Brahmaputra river system: weathering processes and fluxes of the Bay of Bengal Geochim et Cosmochim. Acta. 53 (5), 997-1010.

Schreiber, B.C., 1998. Sedimentary and diagenetic transitions between carbonates and evaporites, 15th international sedimentological congress Alicante, 704-705.

Schelske, C.L., 1988, Historic Trends in lake Michigan Silica concentrations. International Revue der Gesamten, Hydrobio, 73, 559-591.

Schink, D.R. and Guinasso, N.L., Jr. 1977. Effects of bioturbation on sediments-seawater interactions. Marine Geol., 23, 133-154.

Sharma, C., 1970. Origin of Floating Islands in the lakes at Khajjiar and Rewalsar in Himachal Pradesh. The Paleobotanist 19(3), 270-276.

Sharma, C. 1992. Paleoclimatic oscillations since last deglaciation in western Himalaya: a palynological assay. Palaeobotanist. 40, 374-382.

Sharma, C., Singh, G., 1972. Studies in the lake Quarternary vegetational history in Himachal Pradesh-2 Rewalsar Lake. The Paleobotanist 21(3), 321-338.

Sharma, A.K., Jaiswal, S. Negi, V., Pant, M.C., 1982. Phytoplankton community analysis in the lakes of Kumaun Himalaya. Acta Hydrobiol 93(2), 173-193.

Singh, R., Sharma, V.K., 1999. Geoenvironmental appraisal of Mansar and Surinsar Lakes, Udhampur and Jammu districts. GSI Records 131, 19-24.

Singh, R., Misra, S.M., Shyamananda, R.K., Aggarwal, B.K., Mahajan , I., and Sharma, G. 1986. Morphometry and Catchment Study of Rewalsar Lake, Himachal Pradesh, India with a note on its flora and fauna. Asian Journal of Experimental Sciences, 1, 1-7.

Singh, M., Das, B.K., 1995. Major ion geochemistry of Nainital Lake waters, Kumaun Himalaya, India. Bull. Ind. Geol. Assoc. 28(1):53-63.

Singh, M., Das, B.K., 1996. Major ion chemistry and control of weathering of Bhimtal Lake. Kumaun Himalaya, India. Jour. Env. Scie. China, 8, 3, 367-377.

Singh, R., Hasnain, S.I., 1998. Major ion chemistry and weathering control in a high altitude basin: Alaknanda River, Garhwal Himalaya, India. Jour. Hydrological Sciences, 43, (6), 825-843.

Sinha, R.,Raymahashay, B.C. 2000. Salinity model inferred from two shallow cores at Sambhar Lake, Rajasthan, Jour. Geol., Soc., India, 56, 213-217.

Sinha, R., Raymahashay, B.C., 2004. Evaporite mineralogy and geochemical evolution of the Sambhar Salt Lake, Rajasthan, India. Sed. Geol., 166, 59-71.

Sinha, R., Smykatz-Kloss, W., 2003. Thermal Characterization of lecustrine dolomite from the Sambhar Lake playa, Thar Desert, India, Jour. Therm. Anal. Calorin. 71, 739-750.

Singh, M., 1994, Environmental Geochemical Study of Pollution in lakes of Nainital district Kumaun Himalaya, India, Unpublished Ph.D. thesis, Panjab University, Chandigarh.

Spiker, E.C., and Hatcher. P.G. 1984. Carbon isotope fractionation of Sapropelic organic matter during early diagenesis. Org. Geochem. 5, 283-290.

Sirocko, F., Garbe-Schonberg, D., Montyre, A. and Moltino, B., 1993. Century-scale events in monsoonal climate over the past 24,000 years. Nature, 364, 322-324.

Sirocko, F., Garbe-Schonberg, D., McIntyre A and Moltino, B. 1996. Teleconnections between the subtropical monsoons and high latitude climate during the last deglaciation. Science. 272, 526-529.

Stallard, R.F., 1980. Major element geochemistry of the Amazon I. Precipitation chemistry and the marine contribution to the dissolved load at the time of peak discharge. Jour. Geophys. Res. 86, 9844-9858.

Stallard, R.F., Edmond, J.M., 1983. Geochemistry of Amazon.2 The influence of geology and weathering environment on the dissolved load. Jour. Geophys. Res. 88, 9671-9688.

Stallard, R.F., Edmond, J. M., 1987. Geochemistry of the Amazon. 3. Weathering Chemistry and limits to dissolved inputs. Jour. Geophys. Res. 92, 8293-8302.

Stumm, W., 1992. Chemistry of solid-water-interface Wiley, New York.

Stumm, W., and Morgan, J.J., 1981. Aquatic Chemistry. John Wiley, New York. 780p.

Subramanian, V., 2000. Water, quantity-quality perspective in South Asia. Kingston International Publishers, Surrey, U.K.,256 pp.

Tandon, S.K., 1972. Heavy Minerals and quartz axial ratios as provenance indicators in the Siwalik Sediments around Ramnagar, Kumaun Himalaya, Himalayan Geology 2, 206-221.

Talbot, M.R, and Johnston,T., 1992. A high resolution palaeoclimatic record for the last 27,500 years in tropical West Africa from the carbon and nitrogen isotopic compositon of lacustrine organic matter. Earth Planet. Sci. Lett. 110. 23-37.

Taylor, S.R., McLennan, S.M., 1981. The composition and evolution of the continental crust: rare earth element evidence from sedimentary rocks. Phil.Trans. Royal Soc. London A 301, 381-399.

Taylor, S.R., and McLennan, S.M. 1985. The continental crust: its composition and evolution, Oxford, U.K. Blackwell, London, 312p.

Taylor, S.R., McLennan, S.M., 1995. The geochemical evolution of the continental crust. Rev. Geophys. 33, 241-265.

Trisal, C.L., 1987. Ecology and conservation of Dal Lake Kashmir, Butter-Worth and Co. (Publ.) Ltd.

Tripathi, C., and Safaya, H.L., 1998. The Dal Lake-An Environment problem. Geol. Surv. Ind., Rec., Vol., 113 Pt. B, 47-53.

Valdiya, K.S., 1980. Geology of Kumaon Lesser Himalaya. Wadia Inst. Him. Geol., Dehradun, 1-291.

Valdiya, K.S., 1988. Geology and Natural Environment of Nainital Hills, Kumaun Himalaya. Gyanodaya Prakashan, Nainital. 155pp.

Venkatesh, V., 1965. Geochemical evidence for the origin of the Lonar Crater, Maharashtra, India. Bull. Geol. Soc. Amer. 76, 1315-1316.

Venkatesh, V., 1967. Geology and origin of the Lonar Crater, Maharashtra. Rec. Geol. Surv. India, 97. 30-45.

Vital, H., Stattegger, K., 2000. Major and trace elements of Stream sediments from the lower most Amazon River. Chem. Geol. 168, 151-168.

Vallentyne, J.R., 1974. Algal bowl: Lakes and Man. Ottawa: Environment Canada, 186 pp.

Varadarajan, S. 1974. Prehnite-Pumpellyite-Metagreywacke facies of metamorphism of the metabasites of Bhimtal-Bhowali area, Nainital distt. Kumaun Himalaya, India. Him. Geol. 4, 581-599.

Von Damm K.L., Edmond, J.M., 1984. Reverse weathering in the closed basin lakes of the Ethopian rift in the lake Turkana. Am. J., Sci. 284, 835-862.

Wetzel, R.G., 1975. Limnology, Saunders, Philadelphia,743pp.

Wetzel, R.G., 1983. Limnology, Saunders, Philadelphia, II nd ed. 767pp.

Wildeman, T.R., and Condie, K.C., 1973. Rare earth in Archean graywackes from Wyoming and from the Fig Tree Groups, S. Africa. Geochim. et Cosmochim. Acta. 37, 439-454.

Wildeman, T.R., and Haskin, L.A., 1973. Rare earth's in Precambrian sediments, Geochim. et Cosmochim. Acta. 37, 419-438.

Williams, J.D.N., Jaquet, J.M., Thomas, R.L., 1976. Forms of phosphorus in the surficial sediments of Lake Erie. J. Fish, Res. Can., 33, 413-429.

Wronkiewicz, D.J., Condie. K.C., 1987. Geochemistry of Archean Shales from Withwatersrand Supergroup. South Africa: source area weathering and provenance. Geochim. et Cosmochim. Acta, 51, 2401-2416.

Wronkiewicz, D.J., Condie, K.C., 1989. Geochemistry and provenance of sediments from the Pangola Supergroup, South Africa: evidence for a 3.0 Ga-old Continental Craton. Geochim et Cosmochim, Acta 53, 1537-1549.

Yadav, D.N., 1995, Chemical and isotopic studies in continental areas of the Arabian Sea and Saline lakes of Rajasthan (India). Unpub.Ph.D. thesis, M.S. university, Baroda.

Yadav. D.N., 1997 oxygen isotope study of evaporating brines in Sambhar Lake, Rajasthan (India). Chem. Geol.,138, 109-118.

Young, J.D., 1982. Late Cenozoic Geology of the lower Rio Puerco Valencia and Scorro counties, New Maxico: M.S. thesis New Maxico Institute of Mining and Technology, Socorro, 126p.

Zhang, J., Huang, W.W., Letolle, R., Jusserand, C., 1995. Major element chemistry of the Huanghe (Yellow River), China-weathering processes and chemical fluxes. Jour. Hydro. 168, 173-203.

Zutshi, D.P., 1991. Limnology of high altitude lakes of Himalayan region. Verh. Internat. Verein. Limnol. 24, 1077-1080.

Zutshi, D.P., 1985.The Himalayan Lake ecosystem In: J.S. Singh (Ed.) Environmental Regeneration in Himalaya: concept and strategies. The central

Himalayan Environmental Association and Gyanodaya Prakashan, Nainital 325-342.

Zutshi, D.P. and Gopal, B. 2000. Himalayan Lake ecosystems: current issues and threat perceptions. Verh. Internat. Verein. Limnol. 27,2167-2170.

Zutshi, D.P., Subla, B.A., Khan, M.A., and Wanganeo, A., 1980. Comparative limnology of nine lakes of Jammu & Kashmir Himalaya, Hydrobiologia, 80, 101-112.

□□□

INDEX

□□□